# Sustainable Access to Energy in the Global South

Silvia Hostettler · Ashok Gadgil
Eileen Hazboun
Editors

# Sustainable Access to Energy in the Global South

## Essential Technologies and Implementation Approaches

United Nations
Educational, Scientific and
Cultural Organization

UNESCO Chair in
technologies for development
Lausanne (Switzerland)

ÉCOLE POLYTECHNIQUE
FÉDÉRALE DE LAUSANNE

Springer

*Editors*
Silvia Hostettler
Cooperation & Development Center (CODEV)
Ecole Polytechnique Fédérale de Lausanne
Lausanne
Switzerland

Eileen Hazboun
Cooperation & Development Center (CODEV)
Ecole Polytechnique Fédérale de Lausanne
Lausanne
Switzerland

Ashok Gadgil
Lawrence Berkeley National Laboratory
University of California Berkeley
Berkeley, CA
USA

ISBN 978-3-319-20208-2 ISBN 978-3-319-20209-9 (eBook)
DOI 10.1007/978-3-319-20209-9

Library of Congress Control Number: 2015942481

Springer Cham Heidelberg New York Dordrecht London


Printed on acid-free paper

Springer International Publishing AG Switzerland is part of Springer Science+Business Media (www.springer.com)

# Preface

When I was in elementary school in the 1960s, my younger brother and I spent our summers with our grandparents in our ancestral village. Coming from the city life of Mumbai, with electricity, piped running water, and cooking gas delivered reliably to my parents' apartment, the village was truly a different world. Summertime on the Indian coast is very hot and humid. The rhythm of life was very different from that in Mumbai. In the village, children and adults would hike a kilometer up a hill each day to fetch drinking water. My grandmother would cook all our meals with fuel wood in the mysteriously dark smoky kitchen. At night, the house remained mostly dark despite a small kerosene lantern that was lit for only a couple of hours after sunset. Every few days, my grandmother would grind grain by hand on a heavy stone wheel to make flour for chapatis. There were no phones, television, or radio at that time. Change came slowly over the decades: electricity arrived, real incomes grew, kerosene (and eventually bottled gas) for cooking became affordable, piped water started getting delivered to the households, and modest amenities such as electric fans, fluorescent lights, and radios eventually became routine in the households.

For the past thirty-five years, I have experienced two worlds almost concurrently: the challenges of the developing world and the amenities of life in the industrial north. As researcher at Berkeley, my work on energy efficiency, safe drinking water, and clean cooking often takes me to poor parts of the developing world. These exposures remind me of my own memories of how it feels to live in the dark after sunset. Such memories have informed my approach to solving problems faced by the poorest billions of people on the planet.

Energy access is essential for achieving an acceptable standard of living. The outstanding papers in this book underscore the following three lessons from my own experience:

1. As Peter Drucker remarked, "Ideas do not move mountains—bulldozers move mountains. The ideas show where the bulldozers go to work." Without the mobilization of a great deal of social forces, mountains cannot be moved, and that entails a lot of sweat and dirt, and long, hard, and unglamorous work.

2. It is not just about technology—many complex social interactions with institutions, stated and unstated rules and customs, and social and personal habits and preferences must be taken into account for successful adaption on scale.
3. Imagining that a novel technological gadget alone will solve a problem is short-sighted. It is like suggesting that distributing pencils and paper will make illiterate communities literate. Technology does not operate in a vacuum. It needs trained operators and trainers for the operators. It also needs supply chains of spare parts, skilled maintenance personnel, and a business model that makes each participating stakeholder a winner in some way. For a technology to make a positive impact on scale, we need to think through how we will build capacity and linkages, and provide incentives to all relevant stakeholders. Their success will lead to end users' success.

I wish the reader success in incorporating ideas from this book in their own work.

Berkeley, California
April 2015

Ashok Gadgil

# Acknowledgments

The editors owe a debt of gratitude to the many individuals and organizations who generously contributed their time, insight, and support. First, we would like to thank the members of the scientific committee and our session leaders who guided the conference preparation. They not only made the 2014 EPFL-UNESCO Conference on Technologies for Development (2014 Tech4Dev) a great success, but also laid the foundation for this publication.

We would also like to express our thanks to Prof. Philippe Gillet, vice president for Academic Affairs at the Ecole Polytechnique Fédérale de Lausanne (EPFL), for his presence at the conference and unfailing support to the Cooperation & Development Center (CODEV).

By willingly sharing their considerable expertise and different outlooks, the speakers at the UNESCO Conference brought much food for thought to the table and substantially contributed to its success. Our heartfelt thanks go to Dr. Shashi Buluswar (LIGTT: Lawrence Berkeley National Laboratory), Prof. Karen Scrivener (EPFL), Mr. Anil Sethi (Swiss Extension GmbH), Dr. Jean-Bernard Münch (Swiss Commission for UNESCO), Dr. Jean-Yves Pidoux (City of Lausanne), Dr. Christian Zurbrügg (Eawag: Swiss Federal Institute of Aquatic Science and Technology), and Dr. Jon-Andri Lys (KFPE) for their highly appreciated involvement and support.

Likewise, this project could not have succeeded without the quality and diversity of the contributions of the various authors and researchers. In response to the call for papers, the Scientific Committee evaluated over 140 papers and ultimately selected 125 to be presented at the conference. Of these, 16 were finally chosen based on the following criteria: (1) innovative concept and research questions versus an extension of existing work; (2) originality of the methodology including north–south and south–south partnership; (3) contribution to the discipline as whole; and (4) clarity and understandability. We express our appreciation to all these authors, without whom this publication would not have been possible.

In addition, we would like to very warmly thank Mr. Emmanuel Estoppey and Ms. Jeanne Corthay from the Lavaux UNESCO World Heritage Site who went out of their way to welcome us for the social event. All of the conference participants greatly appreciated the opportunity to spend some time outside the conference halls and to experience the splendor of the Lavaux Vineyards.

Our sincere thanks also go to the team from Ingénieurs du Monde (IDM) and our colleagues at CODEV who contributed extensively to the organization of this conference.

Finally, we are very grateful for the generous patronage of the Swiss Agency for Development and Cooperation (SDC), the Canton de Vaud, the City of Lausanne, the Swiss National Science Foundation (SNSF), Cleantech Alps, and the KFPE—Commission for Research Partnership with Developing Countries, the conference sponsors. Their support and their partnership is critical to the achievement of our common mission which is to identify innovative solutions that are able to reduce poverty and lead the way toward more sustainable development at a global level.

# Scientific Committee and Session Leaders

Pankaj Agarwal, Panitek AG, Liechtenstein
Bipasha Baruah, Western University, Canada
Justin Bishop, University of Cambridge, UK
Jennifer Brant, Innovation Insights, Switzerland
Leo Anthony Celi, Harvard Medical School, Beth Israel Deaconess Medical Center & Massachusetts Institute of Technology, USA
Albrecht Ehrensperger, University of Bern, Switzerland
Marie-Valentine Florin, International Risk Governance Council, Switzerland
Zach Friedman, LIGTT: Institute for Globally Transformative Technologies, USA
Ashok Gadgil, University of California, Berkeley, USA
Mini Govindan, The Energy and Resources Institute, India
Sachiko Hirosue, Ecole Polytechnique Fédérale de Lausanne, Switzerland
Silvia Hostettler, Ecole Polytechnique Fédérale de Lausanne, Switzerland
Tunde Kallai, PASRI—ANPR, Meta Group, TR-Associates Ltd, Switzerland
Prabhu Kandachar, Delft University of Technology, The Netherlands
Walter Karlen, ETH Zurich, Switzerland
Denisa Kera, University of Singapore, Singapore
Bertrand Klaiber, Ecole Polytechnique Fédérale de Lausanne, Switzerland
Papa Amadou Konte, Dakar City Municipality, Senegal
Paula Lytle, The World Bank, USA
Temina Madon, University of California, Berkeley, USA
Charles Martin-Shields, George Mason University, USA
Kinsuk Mitra, InsPIRE Network for Environment, India
François Münger, Swiss Agency for Development and Cooperation, Switzerland
Hung Nguyen-Viet, Hanoi School of Public Health, Vietnam
Vipan Nikore, Cleveland Clinic & Massachusetts Institute of Technology, USA
Ermanno Pietrosemoli, Abdus Salam International Centre for Theoretical Physics, Italy and Fundación "EsLaRed," Venezuela
María Catalina Ramírez, Universidad de los Andes, Colombia
Federico Rosei, University of Quebec, Canada
Hans Schaffers, Aalto University, Finland

Klaus Schönenberger, Ecole Polytechnique Fédérale de Lausanne, Switzerland
Tobias Siegfried, hydrosolutions Ltd., Switzerland
Lucy Stevens, Practical Action, UK
Andrés Felipe Valderrama Pineda, Aalborg University, Denmark
Christian Zurbrügg, Eawag: Swiss Federal Institute of Aquatic Science and Technology, Switzerland

# Contents

## Part III Up-Scaling Energy Solutions



## Part IV Potential of Renewable Energy Technologies

# Editors and Contributors

## About the Editors

**Silvia Hostettler** is the Deputy Director of the Center for Development & Cooperation Center (CODEV) at the Ecole Poyltechnique Fédérale de Lausanne (EPFL). She is responsible for coordinating the research activities at CODEV and for the UNESCO Chair in Technologies for Development. She is also in charge of the educational program offered by CODEV and gives lectures in the field of development cooperation, in particular on the potential of technologies and innovation in the Global South. e-mail: silvia.hostettler@epfl.ch

**Ashok Gadgil** has a doctorate in physics from the University of California, Berkeley (UC Berkeley). He is Area Deputy for Science and Technology for the Energy Technologies Area of Lawrence Berkeley National Laboratory (LBNL), and a Professor of Civil and Environmental Engineering at UC Berkeley. He has substantial experience in technical, economic, and policy research on energy efficiency and its implementation—particularly in developing countries. For example, the utility-sponsored compact fluorescent lamp leasing programs that he pioneered are being successfully implemented by utilities in dozens of emerging-economies and developing countries. Per UNEP estimates, these programs were saving about US$5 billion per year and benefiting about 100 million households by 2011, and the numbers have continued to increase since then. He has several patents and inventions to his credit, among them the "UV Waterworks," a technology to inexpensively disinfect drinking water in the developing countries, which enables offering, at profit, affordable safe drinking water to poor communities in poor countries. In 2014, there were about 5 million daily customers in Asia and Africa for this drinking water. He received the Discover Award in 1996 for the most significant environmental invention of the year, as well as the Popular Science award for "Best of What is New—1996." In recent years, he has worked on ways to inexpensively remove arsenic from Bangladesh drinking water, and on fuel-efficient stoves for Darfur. e-mail: ajgadgil@lbl.gov

**Eileen Hazboun** is the Coordinator for the UNESCO Chair in Technologies for Development at CODEV, EPFL. She is also responsible for the coordination of two postgraduate courses in disaster risk reduction and management of development projects. e-mail: eileen.hazboun@epfl.ch

## Contributors

**Samer Abdelnour** Rotterdam School of Management, Erasmus University, Rotterdam, The Netherlands

**Pankaj Agarwal** Panitek AG, Vaduz, Liechtenstein

**Benjamin Attigah** Deutsche Gesellschaft für Internationale Zusammenarbeit (GIZ) GmbH, Eschborn, Germany

**Bipasha Baruah** Women's Studies, Western University, London, Canada

**Ritu Bharadwaj** Institute for Industrial Productivity, New Delhi, India

**Somnath Bhattacharjee** Institute for Industrial Productivity, New Delhi, India

**Richard E. Blanchard** Centre for Renewable Energy Systems Technology, Loughborough University, Loughborough, UK

**Ewan Bloomfield** Practical Action, London, UK

**Alejandra Boni** Instituto de gestión de la innovación y del conocimiento (INGENIO), Agencia Estatal Consejo Superior de Investigaciones Científicas (CSIC), Universitat Politècnica de València, València, Spain

**Emanuela Colombo** UNESCO Chair in Energy for Sustainable Development, Department of Energy, Politecnico di Milano, Milan, Italy

**Nicolas Crettenand** Centre Universitaire de Recherche sur l'Energie pour la Santé (CURES), Yaoundé, Cameroun; Ecole Polytechnique Fédérale de Lausanne (EPFL), Lausanne, Switzerland

**Sunil Dhingra** The Energy and Resources Institute (TERI), New Delhi, India

**Philip C. Eames** Centre for Renewable Energy Systems Technology, Loughborough University, Loughborough, UK

**Albrecht Ehrensperger** Centre for Development and Environment (CDE), University of Bern, Bern, Switzerland

**Álvaro Fernández-Baldor** Universitat Politècnica de València, València, Spain

**Nicolas Gachet** Université de Lausanne, Lausanne, Switzerland

**Kamau Gachigi** Department of Mechanical and Manufacturing Engineering, University of Nairobi, Nairobi, Kenya

**Michael Gonin** Université de Lausanne, Lausanne, Switzerland

**Mini Govindan** The Energy and Resources Institute (TERI), New Delhi, India

**Silvia Hostettler** Cooperation & Development Center (CODEV), Ecole Polytechnique Fédérale de Lausanne (EPFL), Lausanne, Switzerland

**Abhishek Jain** Council on Energy, Environment and Water, New Delhi, India

**Steve Kariithi Karau** School of Human and Social Sciences, Daystar University, Nairobi, Kenya

**Paul Kattuman** University of Cambridge, Cambridge, UK

**Bertrand Klaiber** EssentialTech, Cooperation & Development Center (CODEV), Ecole Polytechnique Fédérale de Lausanne (EPFL), Lausanne, Switzerland

**Kevin S. Kung** Department of Biological Engineering, Massachusetts Institute of Technology, Cambridge, USA

**Aaron Leopold** Practical Action, London, UK

**Pau Lillo** Universitat Politècnica de València, València, Spain

**Jennifer M. MacLeod** UNESCO Chair in Materials and Technologies for Energy Conversion, Saving and Storage, Institut Nationale de la Recherche Scientifique (INRS), Varennes, Canada

**Stefano Mandelli** UNESCO Chair in Energy for Sustainable Development, Department of Energy, Politecnico di Milano, Milan, Italy

**Lorenzo Mattarolo** UNESCO Chair in Energy for Sustainable Development, Department of Energy, Politecnico di Milano, Milan, Italy

**Lucius Mayer-Tasch** Deutsche Gesellschaft für Internationale Zusammenarbeit (GIZ) GmbH, Eschborn, Germany

**Libby McDonald** Department of Urban Studies and Planning, Massachusetts Institute of Technology, Cambridge, USA

**Amber Meikle** Practical Action, London, UK

**Bilal Mirza** Centre for Policy Studies, COMSATS Institute of Information Technology, Islamabad, Pakistan; United Nations University-MERIT, Maastricht University, Maastricht, The Netherlands

**Kinsuk Mitra** InsPIRE Network for Environment, New Delhi, India

**Ange Mugisha** Lake Kivu Monitoring Program, Energy and Water Sanitation Authority, Kigali, Rwanda

**Alice Muzana** Lake Kivu Monitoring Program, Energy and Water Sanitation Authority, Kigali, Rwanda

**Guy Merlin Ngounou** Centre Universitaire de Recherche sur l'Energie pour la Santé (CURES), Ecole Nationale Supérieure Polytechnique de Yaoundé (ENSPY), Yaoundé, Cameroun

**Natacha Pasche** Limnology Center, Ecole Polytechnique Fédérale de Lausanne, Lausanne, Switzerland; Lake Kivu Monitoring Program, Energy and Water Sanitation Authority, Kigali, Rwanda

**Monika Rammelt** Deutsche Gesellschaft für Internationale Zusammenarbeit (GIZ) GmbH, Eschborn, Germany

**Thorsten Reckerzügl** Caritas Switzerland, Lucerne, Switzerland

**Samuel Wanderi Rigu** Department of Agribusiness Management, Safi Organics, University of Nairobi, Nairobi, Kenya

**Francesco Romeo** UNESCO Chair in Energy for Sustainable Development, Department of Energy, Politecnico di Milano, Milan, Italy

**Federico Rosei** UNESCO Chair in Materials and Technologies for Energy Conversion, Saving and Storage, Institut Nationale de la Recherche Scientifique (INRS), Varennes, Canada

**Edouard Rugema** Lake Kivu Monitoring Program, Energy and Water Sanitation Authority, Kigali, Rwanda

**Temilade Sesan** Centre for Petroleum, Energy Economics and Law, University of Ibadan, Ibadan, Nigeria

**Shirish Sinha** Swiss Agency for Development and Cooperation (SDC), New Delhi, India

**Lucy Stevens** Practical Action, London, UK

**Janvière Tuyisenge** Lake Kivu Monitoring Program, Energy and Water Sanitation Authority, Kigali, Rwanda

**Augusta Umutoni** Lake Kivu Monitoring Program, Energy and Water Sanitation Authority, Kigali, Rwanda

**Aline Uwasempabuka** Lake Kivu Monitoring Program, Energy and Water Sanitation Authority, Kigali, Rwanda

**Susanne Wymann von Dach** Centre for Development and Environment (CDE), University of Bern, Bern, Switzerland

**David R. Walwyn** Department of Engineering and Technology Management, University of Pretoria, Pretoria, South Africa

**Collen Zalengera** Mzuzu University, Luwinga, Malawi

**Daniel Ziegerer** Swiss Agency for Development and Cooperation (SDC), New Delhi, India

# Part I
# Introduction

# Chapter 1
# Energy Challenges in the Global South

**Silvia Hostettler**

**Abstract** Energy needs at the global level are gigantic and steadily increasing while the need to reduce carbon emissions is vital if catastrophic climate change is to be avoided. In the Global South, a large part of the population still lacks access to energy, which is crucial for poverty alleviation via the creation of employment and better health and education systems. A great deal of hope is placed on sustainable energy to connect economic growth to increased social equity while preserving the environment. At the same time, energy efficiency and the share of renewable energies in the global energy mix needs to increase unceasingly. This book discusses the challenges of solving complex social and environmental problems with technology in the Global South. While encouraging technology interventions are presented, the limits of technology and the untapped potential of involving women as key stakeholders for energy initiatives constitute also a key focus of this publication. Providing sustainable, reliable, and affordable energy to all will require successful coordination and integration between sectors, stakeholders, policies, infrastructures, and technologies. Promoting renewables, increasing energy efficiency, determinedly addressing the challenges of tailoring solutions to unique socioeconomic contexts, creating local ownership, and aiming for massive up-scaling remain the key issues.

## 1.1 Introduction

In 2014, 1.3 billion people worldwide still lack access to reliable and affordable energy (IEA 2014). At the same time, energy needs will increase by 45 % within the next 15 years (UN 2012). The global electricity demand is growing almost twice as fast as the total energy consumption—a demand that is difficult to meet, even

S. Hostettler (✉)
Cooperation & Development Center (CODEV), Ecole Polytechnique Fédérale de Lausanne (EPFL), Lausanne, Switzerland
e-mail: silvia.hostettler@epfl.ch



S. Hostettler et al. (eds.), *Sustainable Access to Energy in the Global South*,
DOI 10.1007/978-3-319-20209-9_1

more so in view of the missing and aging power sector infrastructure (IEA 2012). A great deal of hope is placed on sustainable energy to connect economic growth to increased social equity while preserving our natural resources. Electricity can empower communities by enabling children to study after dark and clinics to store vaccines and perform life-saving operations. It can improve access to information, security at night, and support the growth of local businesses. The United Nations (UN) launched the Sustainable Energy for All initiative to build the clean energy economies of the future.[1] However, taking into account all new developments and policies, the world still does not have a more sustainable global energy system. Demand will grow by at least one-third until 2035, with India, China, and the Middle East driving 60 % of the increase (IEA 2012) while the world economy might grow threefold in this time (IEA 2012). It will be a formidable challenge to decarbonize the world energy system while also safeguarding that electricity and energy services are accessible to all. Meeting this objective will require a much faster transition to low-carbon energy than what has been achieved thus far (Sachs 2014).

The UN has declared 2014–2024 the decade of Sustainable Energy for All, inciting governments to renew their commitments toward this goal. However, much more needs to be done. The International Energy Agency (IEA) estimates that by 2030, one billion people will still be without electricity and 2.6 billion without clean cooking facilities (2012). In sub-Saharan Africa alone, 600 million people lack access to energy, which is crucial for economic growth. For instance, this region represents 13 % of the world population but only 4 % of the global energy demand (IEA 2014). Only seven countries have electricity access rates over 50 %. The remaining countries in Africa have an average grid access rate of just 20 % (Castellano et al. 2015). Sub-Saharan Africa has significant potential power sources such as solar, wind, and hydropower; however, most effort is concentrated on centralized, large-scale, and grid-based approaches, and few resources are dedicated to providing energy to low-income families.

## 1.2 The Need for More Efficiency

The challenges of providing access to reliable and affordable energy to all are manifold, intertwined, and highly complex. Increased energy efficiency has great potential to lower energy needs and reduce carbon emissions. It is therefore fundamental that policy makers from different sectors integrate energy efficiency as a key measure to positively influence global energy and climate trends. There are encouraging signs, such as China targeting a 16 % reduction in energy intensity, the European Union (EU) planning for a 20 % cut in energy demand, and Japan aiming to reduce electricity consumption by 10 % by 2030 (IEA 2012). Policy makers

[1](www.se4all.org) accessed 7 April 2015.

increasingly realize that energy efficiency is critical to ensure energy security and sustainable development. However, despite new policies, a large share of the potential to improve the energy efficiency is linked to the construction sector and remains untapped (IEA 2014). Energy efficiency needs more visibility by promoting the economic gains to be made. This must be supported by making energy efficiency more affordable by creating appropriate business models and incentives. Governments have a decisive role to play in the mainstream implementation of energy-efficient technologies, which is central to reducing the carbon emissions and avoiding catastrophic climate change. Nonetheless, the fact that considerable reserves of fossil fuels are still available hinders the willingness to give sufficient importance to improve energy efficiency. Fossil fuels receive six times more in subsidies than renewable energies (IEA 2012). In emerging economies such as China, India, and countries in the Middle East, growth in oil consumption, very much linked to the transport sector, more than outweighs the reduced demand in the OECD countries, steadily increasing oil use (IEA 2012). At the same time, technological progress leads to high-efficiency, coal-fired technologies, and efficient carbon capture and storage mechanisms might allow for a gradual transition to a low-carbon power system (IEA 2014).

The IEA (2012) defines access to modern energy services as a household having access to electricity and to a relatively clean, safe means of cooking. The IEA also considers consumption of a specified minimum level of electricity in its definition; the amount varies based on whether the household is in a rural or an urban area. The source of the electricity can be a grid, a mini-grid, or from an off-grid electricity generating system. This book on sustainable access to energy in the Global South illuminates many of the challenges of providing energy at the household and community level.

## 1.3 The Potential of Renewables

The importance of renewable energy such as hydro, wind, and solar power is steadily increasing, reinforced by rising fuel prices, falling technology costs, and subsidies (IEA 2012). Emerging and developing countries in particular have the chance to leapfrog to renewable energy sources, jumping the phase of the fixed network and leapfrogging ahead to a flexible system of multiple, interconnected mini-grids. In many emerging countries, renewables already provide the most affordable source of energy for off-grid and mini-grid systems (IRENA 2014). Brazil, for instance, remains one of the least carbon-intensive energy sectors in the world and has become a world leader in renewable energy (IEA 2013). Power generation from renewable sources is growing twice as fast in non-OECD countries and continents, led by China, India, Latin America, and Africa. The largest share of growth in renewable-based power generation at the global level comes from wind power (34 %), followed by hydropower (30 %) and solar technologies (18 %) (IEA 2014). Even though this is a very positive sign, the combination and integration of different energy sources from a market and technical point of view is becoming more challenging and requires targeted training and capacity building as suggested by MacLeod and Rosei (Chap. 20). Promising new technologies will only be successfully deployed if the expertise to implement, monitor, and maintain sustainable energy infrastructures exists at the global level. Pache et al. (Chap. 21) report on successful capacity-building strategies on environmental monitoring allowing for sustainable energy extraction in Lake Kivu in Africa.

Creating a renewable-based energy system in developing and emerging economies would be a prodigious leap. Renewable energy technologies such as large-scale hydro, geothermal, wind, and solar energy are already providing 30 % of the world's energy needs (IRENA 2014). Wind and solar power are able to compete with non-renewable energy sources while subsidies are decreased. Solar photovoltaic prices have decreased by 80 % since 2008, and onshore wind electricity has decreased by 18 % (IRENA 2014). Total investment in renewable energy has more than quadrupled, from US$55 billion in 2004 to US$214 billion in 2013. This will not be enough to avert serious climate change, which will require investments of at least US$550 billion per year (IRENA 2014), but it shows that a significant change over a relatively short period of time is possible.

## 1.4 Socioeconomic Benefits of Energy Access

Providing low-income families access to energy is considered central to poverty alleviation. In Chap. 3, Leopold et al. argue that despite a substantial political effort at the global level, low-income families might not be the ones to benefit. In an effort to change this, the nongovernmental organization *Practical Action* has developed an Energy Market System Framework to support the creation and maintenance of sustainable energy markets that enable access for the world's poorest people. However, providing access to energy is not enough. As Attigah et al. describe in Chap. 4, electrification needs to be coupled with productive uses in order to achieve socioeconomic benefits for the target population. In order to evaluate the impact of projects on the livelihoods of the intended beneficiaries, Mattarolo et al. have developed an energy-specific, people-oriented monitoring and evaluation approach (Chap. 5). Reckerzügel describes a social enterprise in Bali that was founded to transform the used cooking oil into biodiesel, simultaneously creating income, substituting fossil fuel, and contributing to climate change mitigation (Chap. 22). This project, based on an integrated approach, has the potential to become a best practice for used cooking oil transformation in emerging market countries. This will require persuasive political support to create a legal framework oriented toward recycling, a sustainable business model, and continued financial support for further capacity building. Sinha et al. (Chap. 15) discuss the complexity but also the opportunities such projects offer in the context of the large-scale diffusion of biomass thermal gasifiers in India. Klaiber (Chap. 2) argues for integrated approaches that build profitability along the whole value chain, including the development, implementation, exploitation, and maintenance of appropriate technologies leading to energy security.

## 1.5 Upscaling

Successful upscaling and ensuring long-term sustainability of improved energy solutions remains at the same time a key challenge and a key necessity. Many improved energy projects are overstating the technical solution and do not survive the end of the financial support ensured during their projects' duration. Others never go beyond the successful prototype stage. Yet others are met with resistance by local governments, as was the case with solar photovoltaics (PV) in South Africa despite the fact that PV technology is able to stimulate economic growth, create employment, and reduce inequalities, as Walwyn's study shows (Chap. 13). Ehrensperger and Wymann von Dach (Chap. 7) discuss several key obstacles to long-term sustainability related to project design, management, implementation, and stakeholder participation. For instance, a case study by Mirza (Chap. 10) on Pakistani solar PV systems illustrates how an exclusive focus on technological aspects that neglects a thorough analysis of the energy needs in the community may result in low local ownership of the project. Similarly, local power dynamics and inappropriate subsidies can hinder project implementation and upscaling. Zalengera et al. (Chap. 9) found that while main grid electricity was too expensive for households on Likoma Island in Malawi, the proposed cheaper energy alternatives were not adapted to local energy needs. This illustrates once again the unavoidable trade-offs that need to be made between tailoring energy solutions to specific local contexts and identifying solutions that can be massively upscaled to the regional or even national level. Certain authors, such as Ngounou et al. (Chap. 6), are convinced that there are no standard solutions and that the key to success lies in developing holistic approaches. In their study of decentralized rural electrification systems, Jain and Kattuman (Chap. 11) share the same view and argue that proper management of socioeconomic, operational, environmental, and economic challenges remains crucial. Based on their findings, they recommend that the transition from success stories to upscaling should be based on standardizing the planning approach while customizing the solution.

## 1.6 Gender-Blind Technology

Much hope is placed on technology to solve the worlds' most pressing problems; it is very tempting to overstate the potential of technical solutions. However, beside technology, taking the socioeconomic dynamics of a given context into account will be equally if not more important regarding the success of an intervention (Baruah 2015). Technology by itself will not be able to solve complex social and environmental problems. Baruah and Govindan (Chap. 16) point out that women have less access to new technologies almost everywhere in the world. Nonetheless, the gender implications of promoting new technologies, in particular in the transition to low-carbon economies, remain largely unquestioned. Baruah and Govindan highlight the fact women in science, technology, and energy hold few ministerial positions, and women are rarely considered to be key stakeholders for energy initiatives. They conclude that gender-blind projects and programs by governments, civil society, private sector, or international aid organizations still tend to marginalize women. Fernández-Baldor et al. (Chap. 17) show in their study in Peru how even well-intended technical interventions can reinforce traditional gender roles and entrench intra-household gender inequalities if they do not take into account the fact that technology can contribute unequally to the expansion of people's capabilities. However, gender inclusiveness in the renewable energy sector has considerable potential to support women's empowerment, which technological innovation alone does not guarantee. Skutch (2005) argues that the manner in which an energy service is planned, implemented, and maintained might yield even more positive effects for women than the technology itself.

It is therefore crucial to foster constructive exchange regarding the limits of technology, particularly with respect to the complex problems they are supposed to solve, as Abdelnour points out in Chap. 18. At the same time, using women's empowerment as a pretext and marketing/advocacy platform without rigorously addressing gender inequalities lacks the needed effectiveness in attaining the goal of poverty alleviation. "The unwillingness to fundamentally question the suitability to solve complex problems enables their repackaging as solutions to impending crisis" (Abdelnour 2011).

## 1.7 Conclusion

Providing sustainable, reliable, and affordable energy to all will require successful coordination and integration between sectors, stakeholders, policies, infrastructures, and technologies. Current climate scenarios predict a long-term average global temperature increase of 3.6 °C (IEA 2014). This will lead to a range of very alarming global changes such as rising sea levels, species extinction, food production crises, and an increase in storms and droughts. However, there are also some positive developments. For instance, the costs of solar PV fell by two-thirds between the end of 2009 and 2013 which has been compared to the speed of change in the information technology revolution. In Denmark, wind has become the cheapest energy source of all. In Germany, almost half of all renewable generation is now owned by households and farmers, marking a profound shift in control (IRENA 2014). Progress in energy efficiency and technology gives some cause for hope, as long as political efforts are sustained—a combination of multiple technologies might offer the most potential. Renewable energy is increasingly seen as the affordable solution to meeting the world's rising energy demand while mitigating climate change. For instance, Bharadwaj and Bhattacharjee (Chap. 14) investigated the potential of renewable energy-based mini-grids for rural electrification in India and Africa. Many green mini-grid projects have successfully dealt with the challenges of large-scale deployment and led to effective poverty reduction with the help of smart decision tools.

A sustainable future requires making the best choices aiming simultaneously for energy access for all while achieving economic and environmental objectives. As can be seen in the many chapters of this book, renewable energy also has the potential to provide communities currently left off the grid with electric power, opening up a range of new opportunities for socioeconomic development in the fields of education, health, and access to information and communication technology. The International Renewable Energy Agency (IEA 2014) puts forward the most optimistic scenarios, estimating that doubling the share of renewables in the global energy mix, reinforced by greater energy efficiency, could keep atmospheric carbon below the 450 ppm and thereby below the level at which catastrophic climate change would occur. Let us hope that this is indeed still possible. Promoting renewables, increasing energy efficiency, and determinedly addressing the challenges of tailoring solutions to unique socioeconomic contexts while at the same time aiming for massive upscaling, creating local ownership, and including women in posts of responsibility are the key issues to address.

## References

Abdelnour, S. (2011). Repackaging the efficient stove as the 'solution' to crises in Darfur. *Boiling Point 59*. http://www.hedon.info/BP59_From+Fuelwood+to+GBV. Accessed April 19, 2015.

Baruah, B. (2015). Creating opportunities for women in the renewable energy sector: Findings from research in India. *Feminist Economics, 21*(2), 53–76.

Castellano, A., Kendall, A., Nikomarov, M., & Swemmer, T. (2015). Brighter Africa: The growth potential of the sub-Saharan electricity sector. McKinsey & Company. http://www.mckinsey.com/insights/energy_resources_materials/powering_africa. Accessed April 7, 2015.

International Energy Agency [IEA]. (2012). *World Energy Outlook 2012*. Paris: OECD/International Energy Agency Publications.

International Energy Agency [IEA]. (2013). *World Energy Outlook 2013*. Paris: OECD/International Energy Agency Publications.

International Energy Agency [IEA]. (2014). *World Energy Outlook 2014*. Paris: OECD/International Energy Agency Publications.

International Renewable Energy Agency [IRENA]. (2014). REthinking energy: Towards a new power system. http://www.irena.org/rethinking/Rethinking_FullReport_web_view.pdf. Accessed April 2, 2015.

Sachs, J. D. (2014). *The age of sustainable development*. New York: Columbia University Press.

Skutch, M. M. (2005). Gender analysis for energy projects and programmes. *Energy for Development, 9*(1), 37–52.

United Nations Secretary-General's High-level Panel on Global Sustainability [UN]. (2012). *Resilient people, resilient planet: A future worth choosing*. New York: Columbia University Press.

# Part II
# Socioeconomic Benefits of Energy Access

# Chapter 2
# Holistic and Systemic Approaches to Implement Energy Access Solutions in the Global South

**Bertrand Klaiber**

**Abstract** Energy is an essential resource to improve the quality of human life, and to enable both societal development and economic growth. Universal access to energy requires adopting a comprehensive approach to developing and deploying clean, affordable, and reliable solutions in a sustainable and scalable way. These solutions involve a large range of appropriate technologies from energy production, distribution, and storage, to safe and efficient usage. This comprehensive outlook requires considering the whole life cycle of the products, systems, and services as well as the entire value chain, including all the stakeholders, from global and local players in the private sector to public authorities and civil society. The four contributions of Part II (Chaps. 3–6) describe some key learnings and success factors to provide access to energy to remote or resource-poor communities from a global perspective.

## 2.1 Energy Is Essential for Development

Energy is essential for economic and social development and improved quality of life. Access to affordable and reliable energy is a prerequisite for progress in most sectors, including health care, education, livelihoods, household lighting, cooking or heating services, transportation, and information and communication technologies (ICTs). Although there is a need for a massive increase in energy supply in the Global South, this expansion is inhibited by expensive, underdeveloped, or failing energy infrastructures. In this context, a large range of affordable and appropriate technologies and services are required, from energy production, distribution, and storage to safe and efficient usage.

B. Klaiber (✉)
EssentialTech, Cooperation & Development Center (CODEV), Ecole Polytechnique Fédérale de Lausanne (EPFL), Lausanne, Switzerland
e-mail: bertrand.klaiber@alumni.epfl.ch



S. Hostettler et al. (eds.), *Sustainable Access to Energy in the Global South*,
DOI 10.1007/978-3-319-20209-9_2

## 2.2 Complexity of Energy Access Projects

In the broader sense, energy refers to *natural resources* (oil, gas, uranium, biomass, solar, or wind), *production* or *conversion* (hydroelectric, geothermal, or coal-fired power plants; wind turbines; photovoltaic installations; cogeneration; or waste combustion), *carriers* (fuel or electricity), *distribution* (transportation, pipelines, or power grids), and *consumption* (energy services and use). It is essential to develop or improve the entire ***energy chain*** to provide access to energy. Many different technologies are involved at each level, and these are closely related to financial, political, and social factors.

Energy projects must address technical challenges, while also taking into consideration all the actors, interactions, and roadblocks along the entire ***value chain***. The concept of the value chain consists of all the activities, processes, and operations needed to achieve the development, implementation, exploitation, and maintenance of appropriate technologies for access to energy. The value chain also involves many different actors, such as technology suppliers, product manufacturers, governmental agencies, nongovernmental organizations (NGOs), international organizations, investors, service companies, and various types of consumers. Indeed, end users may have very different energy needs and consumption profiles depending on whether they are households, public services, or any kind of industries or small and medium-sized enterprises (SMEs).

Technologies and strategies need to be converted into products and solutions that can be successfully deployed in the field. However, these products and solutions generally cannot solve energy issues if adequate social initiatives, business models, and policies do not accompany them. Many deployments of technology have failed due to cultural inappropriateness, the lack of local empowerment and competencies for operation and maintenance, and the absence of adequate financial plans to sustain these solutions. From this perspective, any technology should be considered within a comprehensive framework that includes the resources, capacity, and economic models it requires to function. This creates the challenge of evaluating and influencing several factors that are difficult to measure and quantify, including local skills, competencies, awareness, motivations, traditions, cultural specificities, ability, and willingness to pay for energy services. Finally, energy projects should clearly define the expected social, environmental, and economic impacts that can be monitored by means of key indicators and sustained by appropriate procedures.

## 2.3 Energy Security, Quality, and Efficiency

While the lack of energy seems to be the more urgent matter, energy access projects should be evaluated not only in terms of the number of kilowatts they may deliver over time, but also by the socioeconomic benefits they may bring, such as productive uses, more numerous livelihood opportunities, and improved community services. These outcomes rely not only on the quantity of available energy, but also on *energy security*, which the International Energy Agency (IEA) defines as "the uninterrupted availability of energy sources at an affordable price" (IEA, n.d.). In addition, profitable use of energy requires both adequate *power quality* and *energy efficiency*. The first concept refers to the suitability of electric power for consumer devices, which are frequently not working or damaged in developing countries due to perturbations in the electrical grid. The second concept refers to the goal of reducing the amount of energy required for an expected effect or service.

As a consequence of the declining ratio between affordable energy's availability and energy demand, energy efficiency has become an essential factor for growth in industrialized nations. Likewise, there are great opportunities for policy makers and service companies in developing regions to improve energy productivity, cut energy costs, insulate themselves from future energy shocks, and secure a more sustainable development path. According to research on energy productivity in developing countries conducted by the McKinsey Global Institute (MGI) (Farrell and Remes 2009), measures to improve energy efficiency have the potential to "leave demand some 25 % lower in 2020 than it would otherwise have been. That is a reduction larger than total energy consumption in China today."

Energy security, quality, and efficiency are essential aspects of any energy project, requiring services from skilled personnel and companies in addition to appropriate technologies. Similar to the last decade in industrialized countries, the paradigm in developing economies needs to shift from technical products and solutions to energy services.

## 2.4 Holistic and Systemic Approaches

Addressing the complexity, multiplicity, and diversity of problems faced by energy access projects requires holistic and systemic approaches. Several of these approaches are presented in the next four chapters that illustrate the concept elaborated by the *EssentialTech*[1] program of the Ecole Polytechnique Fédérale de Lausanne (EPFL). This program rests on a multidisciplinary methodology addressing three complementary and interlinked aspects of essential technologies for development:

1. Technologies adapted to the local context;
2. Sociopolitical factors in the value chain; and
3. Business models for deployment.

[1]http://essentialtech.epfl.ch.

### 2.4.1 Technologies Adapted to the Local Context

Technologies are required along the entire energy chain and for all services that rely on energy. In particular, it is crucial to develop solutions that take into account the local context and current situation instead of trying to apply strategies from industrial countries that increasingly show their limitations.

To provide significant and durable benefits, energy technologies should be efficient, sustainable, affordable, adapted to the local context, administered by local people, and use local and renewable resources as much as possible. The impact of such technologies on society and the environment is an essential consideration on a local and global scale. Clean energies should be favored to avoid damaging the local environment and contributing to global warming.

Appropriate technology developments must address a variety of challenges, such as high levels of ambient temperature, humidity and dust, low-skill personnel, a lack of reliable infrastructures and spare parts, low purchasing power, poor logistics facilities, and different cultural and social behaviors. Both performance evaluation and the total cost of ownership of a particular solution require consideration of the entire value chain, including equipment, infrastructure, processes, and capacities. Therefore, co-creation and collaboration with local stakeholders that are more knowledgeable of the local context is essential. This process can lead to frugal innovation, which in turn could bring certain advantages to industrialized countries.

### 2.4.2 Sociopolitical Factors in the Value Chain

The impact of a technology is directly dependent on its ability to be deployed, which requires setting up an entire value chain, of which each element must be taken into account in project development. Social and organizational models must be developed to implement the solutions and the incentives or policies fostering sustainable and scalable deployment (e.g., national regulations, tax incentives, tradable carbon credits, or social impact bonds).

The first step involves assessing a situation in terms of material and human resources, competencies, capacities, task-shifting opportunities, regulatory, administrative and institutional frameworks, and societal organization. Cultural specificities should also be taken into consideration, such as the perception of time and priorities or the sense of ownership, moral values, and accountability. Then, the gap between available and required deployment prerequisites should be addressed. This concerns developing teaching and training programs, appropriate policies, awareness campaigns, new processes and organizations, individual and collective empowerment, alliances with private, public, or civil institutions, and so on.

Then, the challenge is to encourage the different actors to participate in the project and initiate the required structural changes. It is therefore crucial to have an interdisciplinary and multicultural approach in energy projects. This approach does not only concern the project team, but should also be implemented further upstream when designing the technology or elaborating a capacity building program. Implementing this strategy can be challenging in a society that segments and splits everything up, e.g., ministries are often not collaborating, and academic branches and administrative services are frequently very specialized and focused.

### *2.4.3 Business Models for Deployment*

New technologies resulting in revenue streams for business ventures and local players are more likely to generate a long-lasting impact than donations, and also have significant potential for scale-up. Products and services associated with a commercial value are often better delivered, operated, and maintained. Energy projects should therefore consider achieving profitability along the entire value chain in order to encourage all the stakeholders to take ownership of the solutions, and to foster both their sustainability and scalability.

From this perspective, appropriate business models must be developed for the entire deployment of energy solutions and services. Business plans should propose ways to fund the initial costs (e.g., acquisition of equipment and creation of dedicated structures), and define models to cover the operational expenditures and generate revenue streams. Initiatives are required to foster the creation of new business ventures and improve their chances of success. This includes, but is not limited to, start-up incubators and "FabLabs," entrepreneurial mentoring, dissemination of a culture of risk and failure acceptance, and proper supportive plans and policies for SMEs.

Markets may be of very different sizes and types, ranging from high-volume small solutions, such as solar lamps or cook stoves, to single-shot large projects, such as production plants that are generally controlled and financed by governments and international companies. Between those, there is an important yet largely untapped market for community-scale energy solutions, such as rural villages, public buildings, and SMEs. The challenge for these projects is to access important capital investments that are often less structured than micro-credits and national funds.

Both fully subsidized and fully liberal approaches have shown some limits regarding the sustainability of energy access projects. Alternatively, *social entrepreneurship*[2] (Kayser and Budinich 2015) generates financial returns together with social and environmental impact. It requires *venture philanthropy* or *impact*

[2]The social enterprise combines commercial and social goals, with emphasis on the latter. It makes use of traditional venture capital model, based on simple risk and reward, while consciously striving to maximize its social impact. The concept of "Scaling Up Business Solutions to Social Problems" is described more thoroughly in Kayser and Budinich (2015), with several examples in the energy sector.

*investments*[3] (Rodin and Brandenburg 2014), to achieve sustainable deployment of energy solutions.

Another recognized sustainability factor is the implication or empowerment of local communities (e.g., village populations or local workers) to ensure that they take ownership of the solution and assume functioning of the equipment—while avoiding issues that can occur when the responsibility is left to an individual or a small group. It is therefore essential to involve them, as well as other stakeholders and key opinion leaders (KOL), in the development of the technology and the implementation of the solutions. In addition, involvement of the private sector for marketing, industrialization, and distribution can bring significant advantages for deployment, such as experience and scale effects.

## 2.5 Universal Access to Energy

*Universal access to energy* is an extremely complex and ambitious goal due to the multiple technologies, stakeholders, and sectors that are simultaneously involved in energy programs, infrastructures, policies, trade, and services. There is no single universal solution, and thus, energy access projects should take their local context into account. At the same time, these projects rely on more global factors, such as national and international regulations, the prices of resources and equipment, scientific progress, and global warming. Consequently, access to energy can only be achieved through integrated approaches to developing and deploying clean, affordable, and reliable solutions in a sustainable and scalable way. Subsequent dissemination of energy technologies requires identification of generic approaches, global knowledge, and replicable solutions that can be adapted to multiple contexts.

In this respect, the following contributions (Chaps. 3–6) present different strategies and models supported by important lessons from practical experiences. Although each paper is different, they all recognize the importance of a holistic, comprehensive approach to address the multiple problems related to energy access. They describe theoretical frameworks and practical tools for analyzing energy ecosystems and managing energy projects. Each paper has been selected to highlight particular frameworks and key success factors from case studies that could be generalized to provide access to energy in remote or resource-poor communities from a global perspective.

[3]In *The Power of Impact Investing*, Rockefeller Foundation President Judith Rodin and Margot Brandenburg (2014) explain what impact investing is, how it compares to philanthropy and traditional investments, where opportunities are evolving around the world, and how to get started.

## References

Farrell, D., & Remes, J. (2009). Promoting energy efficiency in the developing world. *McKinsey Quarterly*. http://www.mckinsey.com/insights/economic_studies/promoting_energy_efficiency_in_the_developing_world. Accessed 16 January 2015.

International Energy Agency [IEA] (n.d.). *What is energy security?* http://www.iea.org/topics/energysecurity/subtopics/whatisenergysecurity/. Accessed 16 January 2015.

Kayser, O., & Budinich, M. V. (2015). *Scaling up business solutions to social problems: A practical guide for social and corporate entrepreneurs*. London: Palgrave Macmillan.

Rodin, J., & Brandenburg, M. (2014). *The power of impact investing. Putting markets to work for profit and global good*. Philadelphia: Wharton Digital Press.

# Chapter 3
# Toward Universal Energy Access: The Energy Market System Framework

**Aaron Leopold, Ewan Bloomfield, Amber Meikle and Lucy Stevens**

**Abstract** Building on the concept of the energy access ecosystem, this paper presents the Energy Market System Framework recently developed by Practical Action to facilitate sustainable energy market creation for improved access to modern energy services in areas of energy poverty. Energy access and energy poverty have been identified as major global challenges to overcoming global poverty and socioeconomic inequality. These challenges relate not to the unavailability of technology but rather to its inaccessibility. These are therefore issues of technology justice, which lies at core of Practical Action's work. In efforts to address these challenges, the Energy Market System Framework, versions of which have been trialed in Thailand, Malawi, Nepal, and Rwanda, maps and designs specific energy market systems to expand energy access to more and poorer people. The approach begins with an analysis of energy market chains, followed by an analysis of national-level policy, capacity, and finance dimensions. The framework enables identification of pathways for action toward achieving universal energy access and technology justice in energy provision. This paper outlines the functionality of the Energy Market System Framework, which recognizes a range of energy services as well as multiple levels within each energy market system that collectively can deliver total energy access using a mix of energy sources and a range of technologies.

## 3.1 Introduction

In 2010, two in every five people (2.8 billion or 41 % of the world's population) still relied on wood, charcoal, or animal waste to cook their food leading to over four million premature deaths per annum associated with household air pollution. Amazingly, as a cause of death, only high blood pressure, alcohol, and tobacco kill

A. Leopold · E. Bloomfield · A. Meikle · L. Stevens (✉)
Practical Action, London, UK
e-mail: lucy.stevens@practicalaction.org.uk



S. Hostettler et al. (eds.), *Sustainable Access to Energy in the Global South*,
DOI 10.1007/978-3-319-20209-9_3

more people (Banerjee et al. 2013; GACC 2013). Additionally, nearly one in every five people (1.3 billion or 19 %) lived without electricity and one billion are serviced by healthcare centers that lack electricity (IEA 2011). No energy for modern lighting, productive uses, or refrigeration means a major portion of humanity is leading less productive and healthy lives than they could and should be. This lack of clean and modern energy services is not due to a lack of available technical solutions, but rather due to their inaccessibility. Practical Action believes that all people should be able to access the technology to provide energy, that access to energy is core to people's opportunities to develop and live the lives they want to lead, and that a lack of access to energy is a technology injustice.

The current profile of energy access is unprecedented. The United Nations has declared 2014–2024 the decade of Sustainable Energy for All after the successful launching of the UN Secretary-General's Initiative of the same name (SE4ALL). However, commitments to sustainable energy do not automatically translate into delivery for poor communities. Without intervention, it is almost certain that financial flows and private sector investments will replicate those of previous decades: favoring large-scale, grid-based, centralized schemes focusing on increased generation capacity which fail to meet the needs of millions of families living in energy poverty. Recent analysis shows, for example, that of nearly US$7 billion in energy financing by the World Bank in July 2012 to June 2013, only US $243 million went to poverty-targeted off-grid solutions and only 8 % overall went to increasing energy access for the poor (not limited to off-grid) (Oil Change International 2013).

There is a real risk that despite vastly enhanced political attention and funding for energy access, poor people may see little direct improvement in their lives and livelihoods.

Tackling this problem will require a combination of efforts from all stakeholders: government, the private sector, and civil society. We will need to harness the power of the market to help extend the reach of much-needed solutions, but do so through delivery models that value expanding energy access to more and poorer people. Practical Action has developed an Energy Market System Framework to facilitate sustainable energy market creation for improved access to modern energy services.

## 3.2 Defining Energy Access

The IEA defines *access to modern energy services* as 'a household having reliable and affordable access to clean cooking facilities, a first connection to electricity and then an increasing level of electricity consumption over time to reach the regional average' (IEA 2012). It should be noted that this electricity can come from either a grid, mini-grid, or off-grid electricity generating system, rather than the more traditional assumption that electricity only comes through a connection to the grid.

Use of the term 'access to *modern* energy services' is important for emphasizing that people need access to not any type of energy, but particularly forms of energy that are clean and that allow them to use the energy productively and to improve the quality of their lives. In addition, referring to modern energy *services* emphasizes the importance of the end-use technologies which include the appliances that all households require, such as clean burning cooking stoves, electric lightbulbs, and efficient agricultural processing machines. Access to these energy technologies allows households, businesses, and communities, to use energy in the ways they desire to improve their lives in a number of ways.

Practical Action has promoted an expanded understanding of definitions of energy access. We have proposed the concept of *total energy access*, which means households, enterprises, and community services having sufficient access to the full range of energy supplies and services that are required to support human social and economic development (Practical Action 2014). We have proposed a set of minimum standards for energy services. Table 3.1 outlines these at the household level. This approach focuses on energy services rather than supplies by, for example, not specifying minimum electricity consumption levels because the same level of service could be achieved with different amounts of energy consumed.

The IEA and World Bank have now adopted a similar approach in their 'Global Tracking Framework' (Banerjee et al. 2013), which goes a lot further toward identifying the important attributes of energy supplies and the energy services that can be derived from them. Rather than defining access in a binary way (yes or no), it places levels of access on a scale (a set of tiers) which gives a far more realistic picture of the reality of energy access for many poor people. Practical Action has been part of the steering committee shaping this framework, based on our work in

**Table 3.1** Defining total energy access

| Energy service | Minimum standard |
|---|---|
| Lighting | 300 lm for a minimum of four hours per night at household level |
| Cooking and water heating | 1-kg wood fuel or 0.3-kg charcoal or 0.04 kg liquefied petroleum gas (LPG) or 0.2 l of kerosene or ethanol per person per day, taking less than 30 min per household per day to obtain |
| | Minimum efficiency of improved solid fuel stoves to be 40 % greater than a three-stone fire in terms of fuel use |
| | Annual mean concentrations of particulate matter (PM2.5) $<10\ \mu g/m^3$ in households, with interim goals of 15 $\mu g/m^3$ and 35 $\mu g/m^3$ |
| Space heating | Minimum daytime indoor air temperature of 18 °C |
| Cooling | Households can extend life of perishable products by a minimum of 50 % over that allowed by ambient storage |
| | Maximum apparent indoor air temperature of 30 °C |
| Information and communications | People can communicate electronic information from their household |
| | People can access electronic media relevant to their lives and livelihoods in their household |

the Poor People's Energy Outlook. The 2015 Global Tracking Framework report will include frameworks not only for households (electricity and cooking) but also for productive uses and community services.

## 3.3 Energy Delivery Models

Improving energy access requires ensuring people have not only an energy supply, but also the energy *services* that result from that (lighting, heating, etc.). These need to be delivered to the places where end users are and in the required quantities, qualities, and prices. The systems by which energy services are delivered are often referred to as energy delivery models (Bellanca et al. 2013).

Energy delivery models can be defined as systems that include a set of processes and activities aimed at delivering energy services to end users through a market chain; a range of external services that support the good functioning of the market chain itself; and the overarching set of administrative and physical infrastructures that together with the sociocultural context create the enabling or disabling environment for the market chain to function effectively. An energy delivery model often includes a number of stages from the initial locating and capturing of the energy sources, to their processing and supply and distribution and marketing, including the technological infrastructure and the design and production of the appropriate technologies. It also includes the financing of the various components and management and maintenance systems that ensure that the energy is affordable and can be supplied over long time periods to meet a range of needs.

Each of these processes and activities is carried out by an individual, or groups of organizations from entrepreneurs to multinational companies (the market actors of the original concept), who work together to contribute isolated parts or the full range of activities.

These systems are often categorized according to whether they are on-grid, mini-grid, or off-grid:

**On-grid**: Centralized energy generation and distribution at a large scale (generally serving many thousands of people), usually referring to electricity and gas national, regional, and district distribution grids. On-grid system capacities are generally greater than 3 megawatt (MW).

**Mini-grid**: Localized, small- or medium-scale power source (typically between 100 kilowatt (kW) and 3 MW) supplying a local distribution grid connected to several domestic, business, and institutional customers in the locality.

**Off-grid**: Stand-alone energy systems that serve individual domestic, business, and institutional customers.

## 3.4 Energy Access Ecosystem

As we have seen, poor people need access to a range of energy services, delivered through a set of energy delivery models. There needs to be an enabling environment at the national level which can facilitate the growth and increasing reach of these systems. To describe the way in which these systems could be mutually reinforcing within a wider enabling environment, Practical Action has adopted the term 'energy access ecosystem.' This borrows from 'business ecosystems' used in management theory. At its heart is an analogy to a natural ecosystem, recognizing coevolution and collaboration, as well as competition, as a means to achieving a healthy system which can provide the range of products and services and involving a range of companies, organizations, and individuals.

The energy access ecosystem considers three main aspects: finance, policies, and capacities. Through a set of nine indicators, it can be used to compare different countries. It recognizes multiple interrelated systems in an energy sector, which collectively deliver energy supplies and appliances using a mix of energy sources and a range of technologies. A healthy ecosystem is one that maintains the sustainability of its vigor, organization, and resilience. In this context, vigor is a measure of the system's activity or production; organization is a measure of the number and diversity of interactions between the components of the system; and resilience is the system's ability to maintain its structure and function in the presence of stress (Mageau et al. 1995).

To try to overcome energy poverty and achieve the United Nation's Goal of Universal Energy Access for All by 2030 as part of SE4ALL, it is necessary to increase energy access, through the development of healthy energy ecosystems. In these systems, expanding access to more and poorer people will be an outcome valued within the system (Practical Action 2013; also see http://practicalaction.org/energy-access-ecosystems).

## 3.5 Energy Market System Assessment Framework

To try and more clearly define the various required components of energy delivery models, Practical Action has been developing an Energy Market System Assessment Framework together with the European Union Energy Initiative–Partnership Dialogue Facility (EUEI–PDF) and trialed in Thailand in the context of small bioenergy markets, Malawi for solar lamps, and Rwanda and Nepal for improved cookstoves (also see Bloomfield 2012 for other examples). The framework has been designed to help decision makers better understand and analyze energy markets, and define support interventions to overcome constraints and take advantage of opportunities, without involving the market actors. First, we outline the framework and then discuss the design of a workshop to use the framework, illustrated with an example from Rwanda.

### 3.5.1 Levels of the Market System Framework

The framework is divided into three levels, the market chain in *Level 1*, the inputs, services, and finance in *Level 2*, and the enabling environment in *Level 3*, as shown in Fig. 3.1.

**Level 1**: The energy market chain is divided into four or five broad functions, which are the most important stages of delivery of a range of energy resources and services. It is important to note that most energy services have only four functions, while mini-grid electrification requires an additional function at the beginning:

**Function 1** Project development (generally only relevant to electricity mini-grids)
**Function 2** Energy generation or appliance manufacture
**Function 3** Energy or appliance distribution
**Function 4** Energy or appliance retail
**Function 5** End-use energy or appliance consumption

Each function is delivered by one, or several, market chain actors, who are the individuals and companies (large and small, formal and informal) who operate within each energy market and, at any one time, own the energy product, service, or

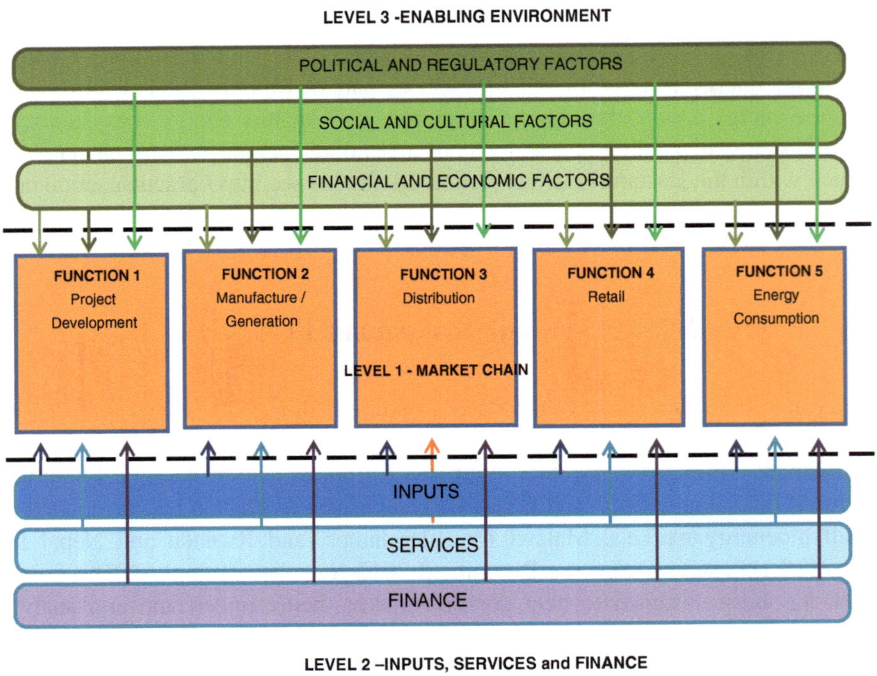


**Fig. 3.1** Simplified summary of an energy market system including the three main levels. *Note* The project development function is only important within mini-grid electrification projects, with the other main energy services only having four broad functions

appliances being delivered. Some market actors only work within one function, while others work across multiple functions. Sometimes, people are hired by a market actor to provide a service, such as a day's worth of labor, but as they do not ever own the product they are not directly included in the market chain, they are instead included within *Level 2* as supporting inputs, services and finance. Typically, each energy product or appliance is transformed in some way as it travels along the market chain. This transformation might be more obvious for physical products such as processed fuels and energy appliances, such as the construction of a solar lantern or a cooking stove, but also includes transformations which are less visible such as regulating voltage around a mini-grid.

It is useful to assess the market actors in each function against the following three criteria. For the overall market to operate effectively and efficiently, there need to be a large enough number of actors, performing well, and who are well connected and coordinated.

- **Number** of actors that in each stage of delivery (current and potential) as well as the geographic distribution of the actors.
- **Performance** of the organizations (how well they operate and how well developed are their business models), as well as the performance of the products.
- **Relationships** between different actors (level of connection, coordination, and competition between actors within each stage and between stages).

**Level 2**: Supporting service actors, who provide the inputs, services, and finance that the market chain actors need to play their role in the market system, for example, with quality, finance, or transport. It is often easier to think about services rather than actors at first and then name the actors behind the services and to divide them into three main groups of inputs, services, and finance.

- Inputs are the wide range of physical resources that each chain actor needs to be able to access to effectively and sustainably deliver their products to the next market actor in the chain.
- Services are needed by each chain actor to better produce, distribute, and sell their products such as design, market research, and quality control and product testing procedures. These are provided by organizations, such as private sector companies, government departments, and community-based organizations.
- Finance includes financing for market chain actors to make products to a high quality and deliver them, and the financial services required by end users to purchase them. These can include loans and equity for the market chain actors and small loans for the end users.

**Level 3**: The enabling environment factors that can be divided into formal (e.g., policies and regulations) and informal factors (cultural norms). It is usually easier to think about issues rather than actors to ensure that all the pertinent constraints are identified. These can be divided into those that can be overcome and those that are typically beyond the control of the market actors such as macroeconomics and global market trends. For those which can be influenced, it is useful to divide into the following three main types of factors: political and regulatory factors, social and cultural factors, and financial and economic factors.

## 3.6 Participatory Market Mapping Workshop (PMMW)

Although the analytical framework is extremely useful in understanding the market system, a participatory process with the market actors themselves is essential to ensure interventions meet the needs of market actors and that the intervention will have their full support. The process itself can help to transform relationships between market actors. In this section, we describe the steps we took in organizing a participatory market mapping workshop (PMMW) on improved cookstoves in Rwanda and summarize the outcomes of the workshop.

### *3.6.1 Preliminary Market Mapping*

Preliminary mapping and analysis is useful to provide an initial overview of a particular energy market and to allow the identification of possible issues and strategies that the market actors may define during the participatory process. It helps build a vision of feasible transformations, assess other interventions already taking place, and identify where gaps exist. The preliminary market mapping also helped us to identify who it would be useful to invite to the workshop (aiming for around 25–30 participants).

### *3.6.2 Supporting Marginalized Market Actors*

It is important to spend time identifying and empowering representatives of less powerful and influential actors to engage in the PMMW. This often includes the producers and marketers of the energy technologies and services. For cookstoves, this included groups such as potters and stove installers. If possible, a meeting should be held with local rural and urban stove producers to introduce and gain understanding and acceptance of the concept and process of the PMMW. This meeting helps in selecting the representative to attend the workshop. This should be someone that others trust and can represent the varied opinions of others without personal biases, can articulate and communicate well with other market actors, and report back and share the key points and findings of the workshop. Some preliminary market mapping can also be done with the producers to generate their key issues/challenges/blockages and possible win-win solutions. Preparing and orienting the beneficiaries for engagement correctly is an important step, to help them grab the opportunity of voicing their opinions and engaging in a change process in their market.

### 3.6.3 Identifying and Hooking the Most Powerful Actors

Getting a broad mix of actors into the workshop is essential to ensure a robust analysis of the market from many perspectives. For powerful groups, we need to find appropriate 'hooks' to attract them. This involves reaching out to the right people and ensuring that the real decision makers are briefed and on board. The anticipated benefits and value for each actor must be communicated explicitly, ideally in face-to-face meetings as well as in writing.

To help with this, we found it useful to establish a small team of champions to lead the process, such as from the local energy authority or a producer association. This also helps to manage expectations. An event promoted by a development organization often generates an expectation of a typical development discussion and not of really doing business between market actors. So is it important not only to identify the potential motivations to attract the actor's attention, but also to think about who should issue the invitation and what would be the most appropriate messaging and languages. It helps to solicit market actors' opinions in terms of how long the event should be and where and when the workshop should take place. For example, an event held at the facility of the government will 'feel' different to one held in an upmarket hotel. A short and snappy half-day event in town may attract more of the busy actors than a two-day event at a resort on the outskirts of town.

### 3.6.4 The Participatory Market Mapping Workshop Event

The objective of the workshop is to cocreate a detailed map of the market system and identify the barriers that exist and opportunities to overcome them. The key outputs are a robust market map and action plan as the basis for market actors to develop their own market system. The map and plan also allows external actors such as nongovernmental organizations (NGOs) and donors to define how they can support market development.

Beyond this, the workshop is *an intervention in its own right*, by increasing access to information and understanding of the market among actors. It brings different actors together who do not usually have an opportunity to interact, facilitating dialogue, trust building, and collaboration. To play this role and ensure everyone can participate equally, it is often necessary to provide translation. This can be a game-changer for the most marginalized actors. Below, we outline some of the sessions used in the PMMW.

**Developing a Market Map**. The participants work together to draw up their market map, looking at the three levels outlined above. However, the map itself is a simplified model and cannot show every nuance and relationship. The discourse between actors, and joint understanding of the market system, is the most relevant output.

It is the facilitator's role to encourage contribution and commitment, help retain interest in the mapping among the group, and steer the group toward a vision for their market by teasing out the causes of market barriers.

**Participatory Market Analysis**. Next, participants use the map to discuss a joint vision of a more functional market system and viable strategies and activities to achieve that vision. This involves identifying opportunities and market failures or blockages, and key leverage points. Leverage points are the points in the system that if touched could change the system as a whole—where the minimum effort could have the maximum effect. Participants need to understand that barriers cannot be removed by any given actor in isolation which is why the problem needs to be addressed in a coordinated and collaborative fashion.

To help with this, participants can think about the market in the future, to imagine how the situation might change in terms of the supply of products to end markets (quantity, quality, variety), as well as the roles and relationships between different actors.

The role of the facilitator needs to be to remind actors of their vision for the market. There can also be a role for an 'expert' who can help challenge dubious solutions and translate what skilled actors mean for the less empowered actors.

**Joint Action Planning**. The workshop now moves from analysis to action. Participants need support to tease out viable interventions. This is done by looking at particular blockages or clusters of blockages. Participants identify win-win solutions/strategies and outline an action plan for each cluster, including desired outcomes, required actions, and suggested dates. Groups are formed around each cluster consisting of the market actors who are most interested in tackling that problem. The clusters represent interest forums where a wide range of strategic, public, and private market actors can now meet, network, build trust, coordinate, and collaborate to address key blockages or exploit key opportunities where they have a stake.

The participatory process ensures cocreation and ownership of the action plan and captures the specific interests and talents of those involved by allowing them to determine which actions should be prioritized. Working collaboratively will better ensure that interventions build on the existing capacities and capabilities of these actors.

**Workshop Wrap-Up**. The workshop wrap-up session serves to emphasize that the workshop was aiming to bring together government officials, NGOs, producers, and other actors to find solutions jointly, rather than approaching these actors independently. It is about market actors getting together to build trust and to find their own ways to unlock the market and start doing business together.

It is important not to raise expectations but to emphasize that it is within the interests of the participants to own their own action to develop their own market.

**Table 3.2** The following supporting service barriers were identified

| | |
|---|---|
| Inputs/physical resources | Biomass fuels, required labor and materials (clay, metal), workshop facilities/equipment/tools/molds, boards and templates, storage space |
| Services | Testing facilities, technical expertise in kiln construction/operation, production training and improved production processes, promotion and marketing of fuels and stoves, metal work skills, carbon credits, access to relevant technology, business skills (planning/bookkeeping/marketing), stove installation, production of molds |
| Finance | Investment financing for producers, end user loans |

**Table 3.3** The following enabling environment factors were defined

| | |
|---|---|
| Political and regulatory factors | Regulation of fuel use, implementation of standards for stove production, specific regulation on stove production and use, cooperation/networking between institutions, research and development |
| Social and cultural factors | High levels of indoor air pollution |
| Financial factors | Taxation on fuel/stoves, carbon financing |

There may be opportunities for development agencies to support specific actions, but this is guided by the drive and vision of the market actors themselves.

After the workshop, outputs should be shared with the participants including the market map (and accompanying intervention matrix). The workshop itself should be just the starting point of collaborative journey to transform the market system, with repeat meetings of all participants, or particular clusters.

### *3.6.5 Outputs from Rwanda Cookstoves PMMW*

A recent participatory market system-mapping workshop was recently carried out in Rwanda to assess the improved cookstove sector of the country. The workshop followed the guidelines illustrated in this paper and identified both barriers and options to overcome them (Tables 3.2 and 3.3).

## 3.7 Conclusion

Practical Action recognizes that the persistent problem of lack of access to clean and modern energy services is one of accessibility not availability. To increase access to a range of services requires new ways of understanding the often complex systems of delivering services to the energy poor and understanding how and what can ensure the necessary change in those systems. It is clear that market systems have already and will continue to play a critical role in delivering energy access, but that currently they are failing to reach the millions of people with the products and services they need. In this context, Practical Action designed the Participatory Energy Market System Development framework to foster the creation and maintenance of sustainable energy markets which deliver on access for the poorest.

## References

Banerjee, S. G., Bhatia, M., Azuela, G. E., Jaques, I., Sarkar, A., Portale, E., et al. (2013). Global tracking framework. Sustainable energy for all. Washington D.C.: The World Bank. http://documents.worldbank.org/curated/en/2013/05/17765643/global-tracking-framework-vol-3-3-main-report. Accessed 16 February 2015.

Bellanca, R., Bloomfield, E., & Kavita, R. (2013). *Delivering energy for development: Models for achieving energy access for the world's poor*. Warwickshire: Practical Action Publishing.

Bloomfield, E. (2011). Energy delivery model tool for understanding and scaling up decentralised energy supply. In *Proceedings of International Conference on Micro Perspectives for Decentralized Energy Supply*, Technische Universität Berlin, 7–8 April.

Bloomfield, E. (2012). Bioenergy market system development: Comparing participatory approaches in Kenya and Sri Lanka. *Boiling Point, 60*, 6–9.

Global Alliance for Clean Cookstoves [GACC]. (2013). *Annual report: Developing markets worldwide 2012–2013*. http://cleancookstoves.org/resources_files/third-annual-report-developing-markets.pdf. Accessed 16 February 2015.

International Energy Agency [IEA]. (2011). Energy for all: Financing access for the poor. Special early excerpt of the World Energy Outlook 2011. http://www.iea.org/publications/freepublications/publication/weo2011_energy_for_all.pdf. Accessed 16 February 2015.

International Energy Agency [IEA]. (2012). World Energy Outlook—methodology for energy access analysis. http://www.worldenergyoutlook.org/media/weowebsite/energymodel/documentation/EnergyAccess_Methodology_2012_FINAL.pdf. Accessed 16 February 2015.

Mageau, M. T., Costanza, R., & Ulanowicz, R. E. (1995). The development, testing, and application of a quantitative assessment of ecosystem health. *Ecosystem Health, 1*(4), 201–213.

Oil Change International. (2013). World Bank Group increases lending for fossil fuels and large hydro: Continues to fail on delivering energy access to the poor. http://priceofoil.org/content/uploads/2013/10/OCI-World-Bank-Energy-Lending-Oct-2013-Final.pdf. Accessed 16 February 2015.

Practical Action (2012). PISCES energy delivery model tool. http://practicalaction.org/energy-delivery-model. Accessed 16 February 2015.

Practical Action. (2012). Poor people's energy outlook. http://practicalaction.org/ppeo2012. Accessed 16 February 2015.

Practical Action. (2013). Poor people's energy outlook. http://practicalaction.org/ppeo2013. Accessed 16 February 2015.

Practical Action. (2014). Poor people's energy outlook. http://practicalaction.org/ppeo2014. Accessed 16 February 2015.

# Chapter 4
# Increasing the Impact of Electrification Through the Promotion of Productive Uses

**Benjamin Attigah, Monika Rammelt and Lucius Mayer-Tasch**

**Abstract** Electrification programs are often justified on the basis of expected socioeconomic benefits resulting from "productive uses" of energy such as increased income and additional employment creation. Productive uses of energy (PUE) are important, not only because they act as the main channel for translating energy access into poverty reduction through increased employment and incomes, but also because of their role in securing the sustainability of electrification. However, public authorities, utilities, and development partners are still not systematically integrating PUE promotion in the design of electrification programs. This can partly be explained by a general lack of knowledge within electric utilities, energy ministries, and energy departments of donor institutions regarding issues not directly related to energy. This paper attempts to start filling the knowledge gap of those in charge of planning and implementing electrification projects about how PUE can be actively promoted. Based on a review of existing experience with PUE promotion, the paper (i) identifies and discusses five different PUE promotion approaches that have been used by governments, utilities, and development organizations and (ii) describes the institutions involved in PUE promotion as well as the roles they typically perform.

## 4.1 Poverty Reduction Through Energy Access?

Investments in electrification are often justified with expected socioeconomic benefits for the target population such as increased income and additional employment creation. However, such economic impacts of electrification only emerge out of productive activities. Hence, efforts to promote productive uses of energy (PUE) are a vital element to translate energy access into positive socio-economic impacts and poverty reduction. Figure 4.1 illustrates how the use of

B. Attigah · M. Rammelt · L. Mayer-Tasch (✉)
Deutsche Gesellschaft für Internationale Zusammenarbeit (GIZ) GmbH, Eschborn, Germany
e-mail: lucius.mayer-tasch@giz.de



S. Hostettler et al. (eds.), *Sustainable Access to Energy in the Global South*,
DOI 10.1007/978-3-319-20209-9_4

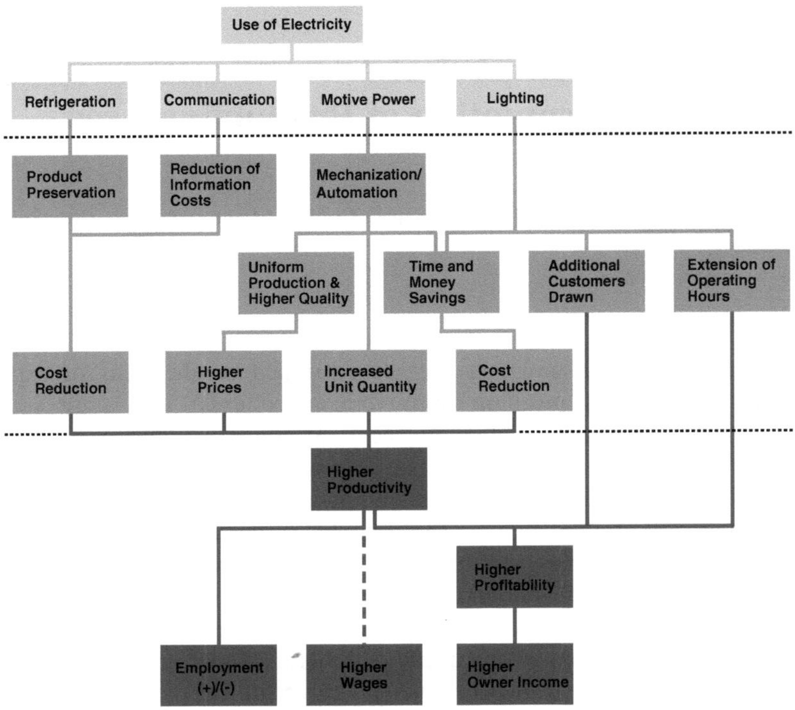


**Fig. 4.1** Pathways from electricity to income generation and employment (Adapted from Peters and Vance 2013)

electricity in enterprises leads to the changes in the income of firm owners and employees as well as employment.

But what exactly are "productive uses of energy"? We can distinguish various different definitions that are currently used in the international debate. Some say PUE are only those activities that directly lead to monetary income, and others have a much broader understanding of the term and include enhancements in welfare through improved health or education. The authors use a fairly narrow definition and understand PUE as "agricultural, commercial, and industrial activities involving energy services as a direct input to the production of goods or provision of services." This can involve the use of various forms of energy including electricity, motive power, and fuels such as biomass or diesel across all economic sectors. PUE can be differentiated from "consumptive use" (i.e., the use of energy services such as household lighting, cooking, and private entertainment) and the use of energy for "community services" (such as health and education). The universal access goal promoted by the Sustainable Energy for All (SE4All) initiative will only be considered achieved if every person on the planet has access to modern energy services for domestic, community, and PUE (ESMAP and IEA 2013).

Productive uses are important, firstly because they act as the main channel for translating energy access into poverty reduction through increased employment and incomes. Secondly, PUE secure the sustainability of energy access programs in two ways: (i) the expected additional income makes the investment in access to modern energy (e.g., investment in a grid connection, electric appliances, or a solar system) viable for energy users who need monetary income in order to be able to afford modern energy services or repay loans for energy-related investments; (ii) energy suppliers need productive uses to achieve a sufficient level of demand and hence revenue that is required for making the construction and operation of electricity infrastructure viable.

Some policy makers and practitioners claim that productive uses happen automatically once access to modern energy is available. However, the PRODUSE study, which assessed the impact of electrification on small and micro-enterprises in three African countries using rigorous quantitative and qualitative evaluation methods, as well as a recent literature review show— that (i) small businesses do not always connect to the grid and that (ii) electricity connection and usage do not necessarily translate into higher firm profits (Mayer-Tasch et al. 2013; Pueyo et al. 2013). In some cases, it could even be observed that enterprises in non-electrified areas fared better than their peers that were connected to the grid. At the same time, the study also showed that electrification can lead to the creation of new firms and attract larger enterprises to newly electrified communities.

It should also be noted that PUE can potentially benefit some enterprises to the detriment of others. This is the case when PUE results in a mere redistribution of wealth within the same community rather than the creation of additional value. For example, an electrified shop may attract new customers at night through better lighting, yet this may also mean that its competitors are losing these customers; in the end, the net effect on income generation in the community may amount to zero. If the designers and implementers of PUE promotion projects want to ensure a positive net effect on incomes, they need to prioritize those end uses that add value to a particular production process.

Despite the importance of PUE for poverty reduction and the viability of energy suppliers, public authorities, utilities, and development partners are still not systematically integrating PUE promotion in the design of rural electrification programs. One reason is the general lack of knowledge (and sometimes interest) within electric utilities, energy ministries, and energy departments of donor institutions regarding issues not directly related to energy. Another reason is the often politically motivated focus on connecting as many people to the grid as possible. This forces the relevant institutions in the energy sector to employ all their financial and human resources to achieve this objective.

## 4.2 How to Promote PUE?

A review of projects from around the globe has shown that productive uses can be promoted in manifold ways. Nevertheless, only few governments and development organizations implementing electrification programs are taking the extra step toward promoting productive uses of the energy services that are provided. Most electrification programs focus solely on the provision of energy to their target groups. This mainly entails interventions related to energy supply: investments in infrastructure (grid lines, transformers, etc.) and technical assistance to energy ministries, rural electrification agencies (REAs), or energy service providers. If at all, the demand side is only considered in the form of awareness programs on health and safety issues, or campaigns aimed at convincing people to be connected to the grid or invest in off-grid systems.

In those cases where energy interventions do promote productive activities, the PUE promotion approach depends on the project setup, context, existing capacities within the project, and partnering institutions. We have reviewed PUE promotion schemes and projects implemented by governments and development organizations and identified the following five different approaches for PUE promotion.

### *4.2.1 Electrification Plus Approach*

Probably, the most common approach to promote PUE is what we call "electrification plus." This approach entails providing a set of complementary services alongside investments in electricity infrastructure. These complementary services include the following:

- Awareness activities and active promotion—e.g., radio/television campaigns and community meetings on productive-use opportunities—to encourage potential and existing businesses to connect to the grid and invest in electric appliances for PUE. This may also include the practical demonstration of PUE appliances.
- Business development services (BDS)—this can entail the identification of the most promising productive uses in the project region and the development of sample business plans; trainings in business management skills and business plan development; activities to strengthen financial literacy (how to open a bank account, how to apply for a loan, etc.); and follow-up meetings with participants of business trainings to assist with the execution of the business plan.
- Improved access to finance—fostering relations with banks, micro-financing institutions or savings cooperatives for suitable loan products to finance PUE appliances, and other related investments.
- Technical advice and training—advice regarding the selection of appropriate high-quality end-use equipment (e.g., energy-efficient electric mills, three-phase welding machines with low power factor, efficient light bulbs, well insulated and robust fridges suitable for bars) and potential equipment suppliers; training in the operation of PUE appliances. This may also include support for converting diesel-powered appliances such as mills to electricity.

- Improved availability of relevant appliances in the project area—for example, by encouraging energy providers or existing shops to offer PUE appliances on a cash or credit basis.

This approach is mainly being used to promote productive use of electricity. However, projects disseminating efficient biomass appliances for productive use such as baking ovens could also include components promoting similar kinds of complementary services.[1]

The *electrification plus approach* has been used and is being used by governments and development organizations (e.g., GIZ, NRECA, UNDP, World Bank) at the micro-, meso-, and macro-level (see, for example, the Afghanistan example described in Box 4.1).

**Box 4.1** Combining Micro-Hydropower Development with Complementary Services in Afghanistan

GIZ's award-winning Renewable Energy Supply for Rural Areas (ESRA) program is supporting the Afghan Ministry of Energy and Water, provincial governments as well as local communities in northern Afghanistan to improve electricity supply and foster local economic development. This includes the development of micro-hydropower (MHP) schemes accompanied by targeted activities for the promotion of PUE. Existing productive uses and PUE potentials are taken into consideration in rural electrification planning and site selection. Once MHP schemes are developed, GIZ supports existing and potential micro-enterprises in newly electrified communities with business trainings and technical advice on the selection, installation, and maintenance of electric equipment. The project further trains local electricians and disseminates information on micro-finance opportunities to local entrepreneurs. New productive uses that have emerged in the electrified villages include carpentry, milling, oil pressing, photocopying, secretarial services, and photography (see also Fig. 4.2). Further support has been provided through the establishment of customer service centers to sustain the benefits to all customers including PUE customers.

### *4.2.2 Call-for-Proposals Approach*

Under the *call-for-proposals approach*, PUE support measures are not offered to all and sundry as can be the case under "Electrification Plus." Rather, measures are specifically targeted toward existing businesses or start-ups by launching a call for proposals. Through the call, businesses are invited to develop and submit a business

[1]For more information on productive use of thermal energy, please refer to Brüderle et al. (2013).

**Fig. 4.2** Carpenter in Faizabad, Afghanistan © GIZ / Oliver Haas

plan for their business ideas. The call for proposals is often held after a workshop or training on PUE and business plan development. Just as with "Electrification Plus," the project implementing agency can support the entrepreneurs by providing complementary services. The *call-for-proposals approach* ensures that support is targeted specifically to those beneficiaries that show initiative, and already possess sufficient entrepreneurial skills to write convincing business plans. Successful examples of PUE applications function as showcases that can stimulate PUE activities on a larger scale in the region.

### *4.2.3 Application-Centered Approach*

The *application-centered approach* is used, as the name indicates, to promote the adoption of PUE appliances such as solar pumps for agricultural irrigation. At the core of the approach lies the specific technology which is to be introduced and disseminated. Hence, the PUE intervention is not necessarily part of an energy project but can be also embedded in a wider program targeting related sectors such as agricultural or local economic development.

Typically, the project implementing agency attempts to create a market or expand the existing market for the selected application. Usually, the first step is to assess the market potential as well as the technical and financial viability of the technology in the local context. This is often followed by the development of a market introduction or market development strategy. Relevant activities include introducing and demonstrating the technology, creating awareness and stimulating demand (e.g., through pilot installations and promotional materials), training technology providers (e.g., in design, installation, and maintenance), and developing an appropriate financing mechanism (which is necessary in most cases as up-front investments are often too high for micro-enterprises).

**Fig. 4.3** Solar irrigation system in Bihar, India © GIZ / Enrico Fabian

It is important for the long-term effectiveness of this approach that the technology providers (producers, technicians, wholesalers, retailers) are working hand in hand with financing institutions and closely watch the needs of the end uses. Financing institutions need to have trust in the technology and the supply chain, because low quality of equipment and installations can constitute a substantial credit risk.

An application that is being promoted by a number of institutions is solar-powered irrigation, as is the case in projects in India, Bangladesh, Benin, and Ghana (Fig. 4.3). Another example is Multi-Functional Platforms (MFP) where a combination of a diesel (or biofuel) engine and productive-use equipment such as a grain mill, an oil press, a water pump, or an alternator is mounted on the same platform, providing the possibility to switch the engine from one machine to the other. In some cases, the alternator is also connected to streetlights or a mini-grid (see also Sect. 4.3.4 on "private sector and civil society").

## 4.2.4 PUE Financing Approach

The *PUE financing approach* focuses on the promotion of selected productive uses through a dedicated financing mechanism. Typically, a fund specifically designed for PUE promotion or a specific loan product is at the core of this PUE promotion approach. The fund can either be application neutral or specifically designed to foster the uptake of a certain application or production process (in this case, it is similar to the *application-centered approach*). The fund or loan product is administered by a local bank or micro-financing institution, which receives financial and/or technical support from the government or development organizations. Together, these institutions develop loan conditions, application procedures as well as the modes of disbursement and loan recovery. Loan applicants are selected based on a comprehensive technical and financial due diligence. In some cases, the loan recipients receive, in addition to the loan, technical backstopping and advice on good business practices. The loan recipients can either be individuals, cooperatives, or community-based organizations. A good example for the *PUE financing approach* is the Electricity End-Use Infrastructure Development (EEUID) scheme of Zimbabwe's Rural Electrification Agency (see Box 4.2).

**Box 4.2** Zimbabwe's Rural Electrification Agency Finances End-Use Appliances

In Zimbabwe, the Rural Electrification Agency (REA) launched the EEUID scheme in 2009, which is a revolving fund extending loans for electrical appliances to SMEs. The fund is administered by a local commercial bank, which is responsible for marketing, due diligence, credit disbursement, and loan recovery. REA has offices in eight provinces of the country, where loan applicants can submit their proposals. Experience with the EEUID scheme has shown that loan conditions must follow a commercial approach to ensure sustainability. If loan conditions are too loose, the scheme may attract a lot of interest from businesses, but this may also have a negative impact on the rate of loan recovery, if the loan recipients are not carefully selected.

### 4.2.5 Cross-Sectoral Approach

In the four approaches described above, promotion of PUE is seen either as an add-on to investments in energy supply or as a financing issue. However, PUE promotion can also be regarded as an integral part of a larger, cross-sectorial intervention. This is the case for many programs aiming at local economic development, small and medium enterprise (SME) promotion, rural development, or agricultural development.

Whereas energy programs quickly reach their limits in terms of resources and capacities for activities that are not related to energy, integrated programs are by design aiming at economic, rural, or agricultural development from the onset. Such programs often go beyond providing the complementary services described in the *electrification plus approach* and follow a *cross-sectoral approach*, for instance, by also providing road infrastructure. Pueyo et al. (2013) argue that integrating electrification with rural development programs is especially relevant for electrification programs operating in deprived areas that do not provide a conducive environment for PUE. For one example of the *cross-sectoral approach* see Box 4.3.

A very comprehensive and complex form of the *cross-sectoral approach* is integrated rural development (IRD). IRD "delivers electricity as part of a wider package of complementary infrastructure, including but not limited to roads, telecommunication, health, and educational facilities" (Kirubi et al. 2009). A number of countries have followed this approach to rural electrification, such as Brazil, Costa Rica, India, and Tunisia. However, the general experience with the IRD approach over more than three decades has raised questions about its efficiency. One main concern is that the approach is too complex and resource intensive. The extra costs for coordinating the interventions of a multitude of stakeholders may therefore outweigh the benefits of cooperation. While cross-sectoral cooperation is essential for the success of PUE promotion, programs also need to limit their scope and the number of cooperation partners in order to keep transaction costs at reasonable levels.

**Box 4.3** Combining Electrification and Local Economic Development in Ghana

One example for the *cross-sectoral approach* is the Program for Sustainable Economic Development (PSED) in Ghana, which is implemented by GIZ together with the Ministry of Trade and Industry and includes an energy component funded by the global *Energising Development (EnDev)* program. PSED supports, among other things, the development of light industrial zones with reliable access to three-phase electricity for productive purposes. PSED works with the local governments of eighteen district capitals that provide the land and the necessary infrastructure such as roads, access to potable water and sanitation for the industrial zones. In addition, the program supports not only the participatory planning and management of the industrial zones but also the provision of BDS and training in environmental management. Figure 4.4 shows a sawing factory operating in one of the light industrial zones.

## 4.3 Who Is Who in PUE Promotion?

For those who are in charge of designing and planning PUE promotion programs, it is important to understand which actors should be involved and what roles they should perform to render the program successful.

### *4.3.1 Ministries and Other Government Institutions*

In some countries, ministries or other government entities have taken the lead in PUE promotion and included a productive-use component into their national electrification program. In such cases, PUE promotion is not limited to pilot projects but can be scaled up and implemented at national level.

**Fig. 4.4** Sawing factory in Ghana © GIZ / Samuel Adoboe

An example of a Ministry of Energy promoting PUE is a program in Peru, where the Directorate General of Rural Electrification (DGER) of the Ministry of Energy and Mines (MEM) is implementing a rural electrification program with a substantial PUE promotion component.[2]

When ministries are promoting PUE, they can agree on cross-sector collaboration at the highest level of government. One potential forum for discussing this is the multi-sector energy committees that have been put in place in Senegal and other mainly west African countries.

Interministerial cooperation is notoriously difficult even in industrialized countries, but there are positive examples such as in the case of India, where the green revolution was promoted in a concerted effort by the Ministries of Agriculture and Energy. In India, the rural electrification program was strongly linked to the promotion of high-yield crop varieties and the spread of irrigated agriculture, facilitated by electric water pumps with subsidized or free electricity.[3]

In almost all cases, governments exert heavy influence on who gets connected when. The example of Bangladesh's rural electrification program shows the great influence of electrification planning on the scope and the type of productive uses that emerge or do not emerge after electrification: This program was initially designed to provide electricity primarily for irrigation and agro-processing and put less emphasis on residential connections or the electrification of social institutions.

[2]Experience with the PUE promotion activities in Peru between 2006 and 2011, which were supported by ESMAP, is documented in Finucane et al. (2012).

[3]Providing electricity to farmers free of charge is not recommended, because of the potential negative effects on the viability of the utility company and the risk of overuse of water for irrigation.

Depending on the degree of decentralization and the human and financial resources they have at their disposal, local governments can also play an important role in PUE promotion.

### *4.3.2 Rural Electrification Agencies*

Following the wave of power market reforms in the 1990s, many governments especially in Sub-Saharan Africa established REAs. In some instances, REAs are engaged in active promotion of PUE. While some REAs focus on ensuring electricity supply to productive end uses, other REAs also support the provision of additional complementary services such as BDS or finance support, which are often delivered by external service providers.

REAs that are promoting PUE can be found in Nepal, Mali, Senegal, Uganda, and Zimbabwe (see also Box 4.2). The Senegalese Rural Electrification Agency (ASER) is financing the energy component of "multi-sectoral energy investment projects" (so-called PREMs, *Programmes Energétiques Multisectoriels* in French) implemented in cooperation with government, donor, or NGO programs. One example of a PREM in northern Senegal involved the construction and electrification of four dairy collection and transformation centers as well as two cattle-feed processing plants.

### *4.3.3 Utilities*

Utilities and energy service providers can also play a crucial role in PUE promotion. Increasing the number of commercial connections and the electricity consumption of productive users helps to generate additional revenues and improve the profitability of the service provider. PUE should therefore be considered as an important element in the marketing strategy of the energy service provider. For small, community-based mini-grid operators that do not have the possibility to cross-subsidize rural residential consumers with the revenue from more profitable areas, achieving a sufficient load factor through PUE is key for long-term sustainability.

One successful example in this context is the aggressive load promotion program of Thailand's public distribution company Provincial Electricity Authority (PEA) that was started in the 1970s and contributed to the success of PEA's rural electrification program.[4] PEA applied a mix of direct and indirect PUE promotion. On the one hand, it created a load promotion unit within its organization that gave direct technical advice to PUE users and carried out comprehensive market research

[4]For more information on this program, see Tuntivate and Barnes (2007).

and awareness raising activities. On the other hand, PEA provided indirect assistance through partnerships with a range of stakeholders, for example, by cooperating with other government agencies to disseminate technical information on potential productive uses to the communities living in the areas of intervention. Most importantly, however, PEA partnered with Thailand's Bank of Agriculture and Agricultural Cooperatives to provide loans to commercial and agricultural end uses (e.g., rice millers) to incentivize them to switch from diesel engines to electric motors.

In Kenya, the state-owned distribution company Kenya Power is promoting PUE through marketing activities that are specifically targeting SMEs. These include pre-electrification customer education on PUE opportunities and partnerships with equipment manufacturers, financial institutions, and other stakeholders. For instance, Kenya Power facilitates the demonstration of PUE equipment by manufacturers at trade fairs, in schools, and in universities. Other examples of utilities that are actively promoting PUE can be found in Brazil, Bangladesh, and Nepal.

Nevertheless, the majority of utilities do not engage in PUE promotion. The complexity of PUE programs which include awareness raising, technical advice, financing, and BDS often makes them appear to be too much effort for too little benefit. On a purely commercial basis, it is less costly and financially much more attractive to extend services for larger industrial and commercial customers instead of focusing on small rural consumers.

When utilities do promote PUE, it is often initiated or supported by development organizations. NRECA has supported productive-use promotion programs of electric cooperatives in a number of countries including Bangladesh, Bolivia, Guatemala, and El Salvador.[5] The EnDev-funded GIZ project in Nepal together with HELVETAS Swiss Intercooperation supports PUE promotion by "Community Rural Electrification Entities" (CREE) that are managing sections of the rural electricity network.

### *4.3.4 Private Sector and Civil Society*

The informal or formal private sector obviously forms the target group of all PUE promotion schemes. Many programs cooperate with industry associations or cooperatives to disseminate information, raise awareness, and provide capacity building for their members.

Some PUE promotion programs are working directly with the private sector or civil society organizations, without public authorities having a central role in implementation. Here, the focus is on providing a power source and at the same time promoting a range of productive end uses.

[5]The NRECA experience with PUE promotion is summarized in NRECA International (n.d.).

For example, the Greening the Tea Industry in East Africa (GTIEA) project works to increase private investments in small hydropower to supply electricity to private tea factories in Kenya. The main goals are to reduce energy costs of the tea industry and provide a clean and reliable source of supply. The GTIEA project supports the tea companies by preparing feasibility studies for small hydropower projects and providing technical assistance during the implementation phase.

Other programs cooperate closely with NGOs or community-based organizations. The UNDP program to disseminate MFPs in West Africa, for example, assisted community-based organizations (mainly women groups) and private entrepreneurs to acquire and operate MFPs. In some cases, they relied on local NGOs to reach out to and train the beneficiaries. In the original concept, the MFPs were owned, operated, and managed by women's groups; however, in practice this model proved problematic, as many of these groups had difficulties in operating and managing the systems. One study on the MFP program found that in the case of Burkina Faso, the majority of village group activities initiated and financially supported by external development actors were not continued when these actors withdrew; the viable businesses (such as milling), however, continued on a private ownership basis (Nygaard 2010).

### *4.3.5 Financing Institutions*

Depending on the approach, most PUE promotion schemes involve financial institutions, most commonly micro-financing institution (MFI) providing credits for investments in access to modern energy (e.g., grid connection and house wiring, solar systems, and efficient biomass appliances) and end-use equipment. These may receive support from the government and development organizations in the form of capital (credit lines, revolving funds) and technical assistance. This could include the development of loan products that match the requirements of businesses that want to invest in electric appliances and machines for PUE.

## 4.4 Conclusion

It is only through the promotion of PUE that energy access interventions can unfold their full potential for economic development and ultimately poverty reduction. However, there is no one size fits all solution to PUE promotion. The existing PUE promotion schemes have applied at least five different approaches for systematically promoting PUE. The selection of the right approach depends on the mandate of the institution that is initiating the program and on the capacities and resources that are available. The approach also depends on whether electrification programs are implemented in areas with a conducive environment for productive uses, such as adequate road and telecommunication infrastructure. If they do not, PUE promotion

**Fig. 4.5** Tailor using electric sowing machine in Senegal © GIZ / Gunnar Wegner

will be significantly more successful if electrification and PUE promotion efforts are integrated with other programs addressing these barriers.

The deeper the PUE promoting activities are rooted within national strategies and institutions, the more likely the PUE promotion can be brought to scale and have a lasting impact. The success of PUE promotion programs depends also on close cooperation and communication of different actors, e.g., different government ministries and/or agencies, utilities, technology providers, financing institutions, NGOs, and development organizations, who should perform roles that are in line with their comparative advantages. However, care should be taken that PUE promotion schemes do not become excessively complex.

Together with ESMAP, the Africa Electrification Initiative, and EUEI PDF, GIZ initiated the PRODUSE portal (www.produse.org), which not only avails a good overview of project examples for PUE promotion around the world, but more importantly offers relevant guides for project implementers on how to go about PUE promotion as well as subsequent monitoring and evaluation exercises. Two studies make up the heart of the portal: the PRODUSE study (Mayer-Tasch et al. 2013), which examined the usage of electricity in micro-enterprises and assessed the effect of electricity usage on the performance of firms in Benin, Ghana, and Uganda, and the PRODUSE manual (Brüderle et al. 2011), a guide for practitioners interested in PUE promotion. PRODUSE is living through the contributions of GIZ and its partners. Hence, comments, additions, and project examples are very welcome! (Figure 4.5).

## References


Brüderle, A., Attigah, B., & Bodenbender, M. (2011). *Productive use of energy (PRODUSE)—A manual for electrification practitioners*. Eschborn: EUEI PDF & GIZ.

Brüderle, A., Diembeck, K., Hartmann, J., Rammelt, M., & Volkmer, H. (2013). *Productive use of thermal energy. An overview of technology options and approaches for promotion*. Eschborn: EUEI PDF & GIZ.

Energy Sector Management Assistance Program [ESMAP] & International Energy Agency [IEA]. (2013). Sustainable energy for all—Global tracking framework. Washington D.C.

Finucane, J., Bogach, V. S., & Garcia, L. E., & Zolezzi, E. H. (2012). Promoting productive uses of electricity in rural areas of Peru: Experience and lessons learned. In *Energy Sector Management Assistance Program (ESMAP)*. Washington DC: World Bank.

Kirubi, C., Jacobson, A., Kammen, D. & Mills, A. (2009). Community-based electric micro-grids can contribute to rural development: Evidence from Kenya. *World Development, 37*(7), 1208 −1221.

Mayer-Tasch, L., Mukherjee, M., & Reiche, K. (Eds.). (2013). *Productive use of energy (PRODUSE): Measuring impacts of electrification on micro-enterprises in Sub-Saharan Africa*. Eschborn: GIZ & ESMAP.

NRECA International (n.d.). Guides for electric cooperative development and rural electrification —Module 9: Productive uses of electricity. Arlington.

Nygaard, N. (2010). Institutional options for rural energy access: Exploring the concept of the multifunctional platform in West Africa. *Energy Policy, 38*(2), 1192–1201.

Peters, J., & Vance, C. (2013). Methodology. In L. Mayer-Tasch, M. Mukherjee, & K. Reiche (Eds.), *Productive use of energy (PRODUSE): Measuring impacts of electrification on micro-enterprises in Sub-Saharan Africa*. Eschborn: GIZ & ESMAP.

Pueyo, A., Gonzalez, F., Dent, C. & DeMartino, S. (2013). The evidence of benefits for poor people of increased renewable electricity generation: Literature review. IDS Evidence Report Nr. 31.

Tuntivate, V., & Barnes, D. (2007). Public distribution and electricity problem solving in rural Thailand. In D. Barnes (Ed.), *Meeting the challenge of rural electrification in developing nations: The experience of successful programs*. Washington DC: Resources for the Future.

# Chapter 5
# An Integrated Monitoring and Evaluation Approach for the Assessment of Energy Development Projects

**Lorenzo Mattarolo, Stefano Mandelli, Francesco Romeo and Emanuela Colombo**

**Abstract** To achieve benefits over time, energy development projects should take into account social aspects that influence their success. To evaluate this success, project monitoring and evaluation (M&E) should not only assess the achievement of expected objectives, but also monitor recipients' roles within the various steps of the project. With regard to this issue, in this paper, the authors describe a new M&E approach that improves on other traditional approaches. The new approach combines the *logical framework approach* (LFA) and the *sustainable livelihoods framework*. While the first represents a widely used tool for assessing the achievements of the project steps, the integration with the latter allows consideration of the importance of the recipients' roles within the project activities in order to induce targeted changes in livelihoods. From this viewpoint, the new integrated M&E approach helps to set up an effective set of indicators to perform a *people-oriented* evaluation of the project aside from a mere evaluation of the completion of the project phases. The proposed integrated M&E has been applied to an Energy Facility project funded by the European Union (EU) and implemented by the Italian nongovernmental organization (NGO) COOPI Cooperazione Internazionale in Malawi.

## 5.1 Monitoring and Evaluation of Energy Development Projects

Modern forms of energy are still not accessible to all, and today, 1.3 billion people live without electricity and 2.6 billion people rely on traditional biomass for cooking, lighting, and heating (IEA 2014; Colombo et al. 2013). From this

L. Mattarolo (✉) · S. Mandelli · F. Romeo · E. Colombo
UNESCO Chair in Energy for Sustainable Development, Department of Energy, Politecnico di Milano, Milan, Italy
e-mail: lorenzo.mattarolo@polimi.it

S. Mandelli
e-mail: stefano.mandelli@polimi.it



S. Hostettler et al. (eds.), *Sustainable Access to Energy in the Global South*,
DOI 10.1007/978-3-319-20209-9_5

perspective, the role that local development projects may play in promoting access to energy is crucial. Such projects primarily address the provision of basic needs for households, but they also represent leverage for enabling productive usage and income-generating activities (European Commission 2003; Peters 2009; Barkat et al. 2002; Bensch et al. 2011).

In order to be successful and to achieve durable benefits over time, energy projects have to consider: (i) the specific features of the targeted context, and (ii) all of the possible consequences that an energy solution may have on social, environmental, and economic dimensions (Brent and Rogers 2010; Cherni et al. 2007). In addition, it is important that project recipients have significant involvement in the various phases of the project and ultimately take full ownership, playing a key role for local changes in behaviors, capabilities, and opportunities (McDonald 1999).

In light of these considerations, a proper monitoring and evaluation (M&E) process of projects should be accomplished. While monitoring usually implies "*a continuing operation conducted by project staff during project implementation to ensure that the project stays on track to achieve its objectives*," evaluation is "*a periodic assessment of the relevance and overall performance of the project [...], commonly done on completion on the project and then some years later in order to assess its longer-term impact*" (Annecke 2008). In the context of development cooperation, M&E is recognized as a fundamental step as it enables international agencies, nongovernmental organization (NGOs), and policymakers to learn from past experiences and to better plan and manage ongoing or future activities (Khander et al. 2009; OECD 1991; Kusek and Rist 2004).

In recent years, the international community has begun a debate on the importance of project evaluation by shifting the focus from the analysis of performed activities and obtained results within the project boundaries and time frame to a more long-term impact on local development (Kremer et al. 2009). In this regard, the Paris Declaration on Aid Effectiveness (OECD 2005) proposed five main points to track implementation and project progress: *ownership, alignment, harmonization, managing for results,* and *mutual accountability. Ownership* and *alignment* especially emphasize the importance of the project recipients with their needs, priorities, and decisions that, in addition to the appropriateness of the implemented technologies, have to be assessed by M&E systems to ensure the long-term effectiveness of development interventions. Therefore, monitoring how a project leads recipients to change the way in which they act, work, or live is also fundamental. This could significantly alter the way a program will assess its performance and impact, since for each achieved step of the project, a correlated change in the target community's behavior should be expected. However, compared to other sectors (such as education, water, or health), where M&E systems are better defined, the energy sector has only recently integrated in the M&E social issues and recipients' perspectives alongside technical and economic factors (Annecke 2008; Khandker et al. 2009; Adams et al. 2006).

With regard to this aspect, in this paper, we introduce a new integrated M&E approach which specifically addresses the monitoring of recipients' roles and the induced livelihood changes through energy development projects.

## 5.2 An Integrated Monitoring and Evaluation Approach

In this section, we describe an integrated monitoring and evaluation (IM&E) approach which aims at assessing the project success as follows: (i) *changes in state* stemming directly from the project activities' results and (ii) people's *changes in behaviors*. In order to match these two viewpoints, the IM&E approach couples the LFA with the sustainable livelihoods framework. While the first represents a management tool for regulating the project steps (detecting the *changes in state*), the latter focuses on improvements in a community's livelihoods (*changes in behaviors*) as the key for project success. Prior to proposing the new approach, we briefly introduce the two frameworks in order to identify their main features.

### 5.2.1 *The Logical Framework Approach*

The LFA is an analytical process developed to assist international development agencies in improving their project planning, management, and evaluation systems (Fry Consultants Inc 1970). The European Commission has adopted the LFA as part of its Project Cycle Management (PCM) system (European Commission 2004). Within the PCM, a set of interlocking concepts have been proposed as part of an overall process which helps structured and systematic analysis of a project idea, so that relevant issues can be considered, criticalities identified, and the objectives reached. Within the analytical process proposed by the LFA, the key operational tool is the logical framework matrix (LFM), which summarizes the key elements of the project plan:

- The project's hierarchy of objectives and results (*project description*);
- How the project's achievements are monitored and evaluated (*indicators* and *sources of verification*); and
- The key external factors critical to the project's success (*assumptions*).

The project description has to be presented according to the *overall objective(s)*, the *specific goal(s)*, the *results,* and the *activities.* Objectively verifiable indicators (OVIs) have to be developed for the M&E of the project's achievement and have to be coupled with related sources of verification. Such indicators have also to be SMART, i.e., *specific*, *measurable*, *available*, *relevant,* and *time-bound.* Lastly, external assumptions are also considered, i.e., those specific conditions and factors outside of the project management's control.

The LFA provides logical links between the four levels of the so-called project *results chain*—i.e., activities, output(s), outcome(s), and impact(s)—which help in the development of the SMART M&E indicators (Danish Energy Management 2009). In this way, the LFA places the project within a broader development context through the distinction between specific goal(s) and overall objective(s) (closer to the long-term *impact* concept). Nevertheless, having its roots in the PCM, the LFA reflects the aim to assist project planning, management, and evaluation, and for the M&E results to be *project oriented* rather than *people oriented* wherever necessary.

### 5.2.2 The Sustainable Livelihoods Framework

The sustainable livelihoods framework (SLF) is a tool to support the implementation of development projects, putting people's livelihoods at the center of expected changes (Scoones 1988; Chambers and Conway 1992; UK DfID 1999). From an operational perspective, the SLF represents an analytical structure to facilitate a broad systematic understanding of the various factors that constrain or enhance livelihood opportunities within a community and to show how they relate to each other (Krantz 2001).

The SLF considers people within a community as operating in a *vulnerability context*, within which they have access to certain *livelihood assets*. Such assets, and the access to them, change and are defined by the social, institutional, and organizational local environment (*transforming structures and processes*). These elements deeply influence the *livelihood strategies* undertaken by people in order to reach their self-defined beneficial *livelihood outcomes*, whose achievement is expected to change the original *livelihoods assets*.

In particular, the *vulnerability context* consists of the environment in which people live. It includes shocks (e.g., natural hazards, economic crises, or conflicts), trends (e.g., demographic or resource trends), products, or employment opportunities, and it represents the part of the framework that lies outside people's control. It has a direct impact upon people's *livelihoods assets,* which include not only the community's material assets, but also the people's capabilities and strategies to use them. These assets are grouped into five community capitals, i.e., natural, physical, human, social, and financial capitals, as is well explained in (Kollmair and Gamper 2002). These *livelihoods assets* and access to them are influenced and shaped by the *transforming structures and processes. Structures* can be described as the hardware (i.e., private and public organizations), while *processes* constitute the software determining the way in which structures and individuals operate and interact (i.e., laws and policies). These three elements determine people's *livelihoods strategies,* i.e., the range and combination of activities and choices that persons undertake in order to achieve their *livelihoods outcomes*, which comprise the *livelihoods assets'* improvement (e.g., more income, increased well-being, and improved food security).

In fact, the SLF places people at the center of changes, and from this perspective, recipients are the key players in a project's success. Indeed, proper strategies capable of achieving project objectives result from people undertaking a proper combination of choices and related activities.

### 5.2.3 The Integrated Approach

Considering what has been highlighted above, developing a more effective M&E approach is necessary to allow comprehensive evaluations which take into account the recipients' perspectives. In this regard, on the one hand, the LFA provides a tool to identify, implement, and evaluate the project steps, but it does not consider recipients' strategies. This leads to the definition of risk and measure indicators oriented toward the projects' boundaries without considering long-term effects on the target community. On the other hand, the SLF represents a tool to investigate how issues concerning recipients link to each other in the development processes; nevertheless, it does not provide an M&E approach. The IM&E introduced and applied in this paper relies on the LFM for the project structure and the M&E modeling, but adds new *people-oriented* elements coming from SLF.

The conceptual structure of the IM&E is introduced in Fig. 5.1. The project *result chain* is linked to the *people-oriented* approach by a set of *people-oriented assumptions,* which are internal assumptions essential to include all along the *result chain, examining* the way in which recipients respond, with their strategies, to the project. These people-oriented assumptions are as follows:

1. *Participation*. The expected results of the project consist of improvements of the livelihoods assets enjoyed by recipients. Since these include material assets as well as people's capabilities to exploit them, beneficiaries' proactive participation is central. Hence, in order to reach the results, the first people-oriented

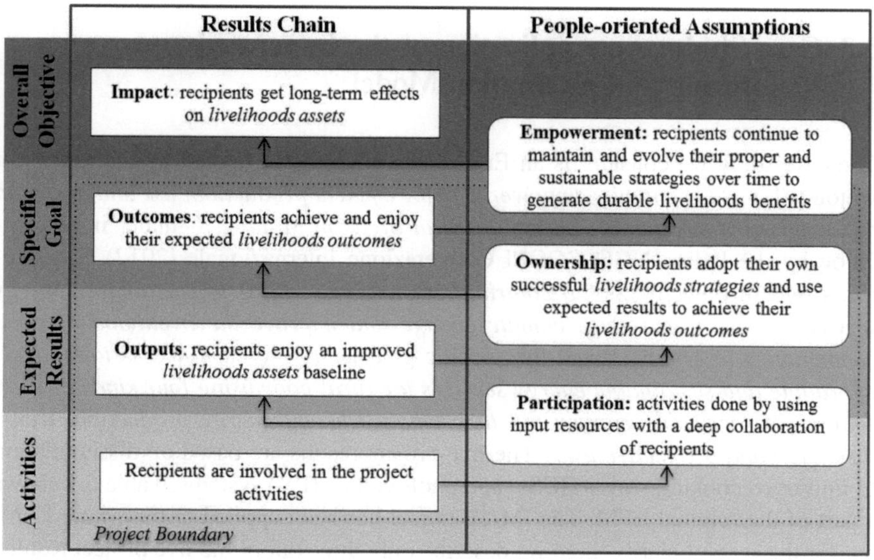


**Fig. 5.1** The new integrated approach

assumption is represented by *participation*, i.e., the extent of collaboration by recipients in initial project activities.

2. *Ownership*. Once the expected results are achieved (i.e., assets have been improved), recipients should take advantage of them and autonomously adopt their own livelihoods strategies to improve their livelihoods. It is thus necessary for recipients to take full ownership of the project. Hence, the second *people-based assumption* is represented by the recipients' *ownership*.
3. *Empowerment*. With the achievement of the specific goal, the project's task is completed, and more long-term benefits for the community may be reached only with the durable adoption of proper strategies, which provide evidence of a long-term *empowerment* of the community.

Introducing the three *people-oriented assumptions* (i.e., participation, ownership, and empowerment) allows us to highlight activities, strategies, and objectives adopted by recipients in response to project activities, and hence to develop more appropriate indicators for the M&E of the project. In line with this new framework, we introduced two modifications to the LFM: (i) project expected results are restructured to better visualize the *people-oriented* components in addition to the *project-oriented* ones, and (ii) *people-oriented* indicators are consequently added at the level of results, specific goals, and overall objectives.

In this way, the IM&E can provide a more explicit framework to monitor the contribution of *people-oriented* components all along the *results chain*, thus improving the LFA framework with the SLF perspective.

## 5.3 Case Study: An Application of the Integrated Monitoring and Evaluation Model

The case study we consider is an Energy Facility project funded by the European Union (EU)—"*Community empowerment for efficient production, use and access to renewable and sustainable energy in rural areas in Malawi*,"—under implementation by the Italian NGO COOPI Cooperazione Internazionale (2012). The project's overall objective is "*to contribute to improve access to sustainable energy services in order to reduce climate change and improve the livelihood of rural communities in Malawi*," and the specific goal is "*to increase access to modern, affordable and sustainable energy services for rural poor using four kinds of local sources of renewable energy and to promote efficient energy production, transformation and productive use*." The five project results are based on dissemination of improved cooking stoves (ICSs) and photovoltaic (PV) systems. Table 5.1 shows a part of the original LFM. The IM&E model has been applied to the whole LFM, but for the sake of clarity, here we describe only three out of the five project results.

**Table 5.1** Considered expected results from the LFM of the COOPI project

| | Project description | Indicators |
|---|---|---|
| R1 | In Kasungu and on Likoma Island, 9000 people will reduce wood consumption by at least 40 % using ICSs for cooking and will produce less smoke | The number of new households using improved cooking facilities (efficient energy saving /fewer smoke emission from stoves) |
| R2 | Six communities in the Kasungu District will use solar energy (nongrid electricity) to increase, by at least 60 %, productivity, income, and work efficiency to irrigate at least 24 new hectares around six existing dams | 1. Number of beneficiaries (700–900 households) accessing nongrid electricity for irrigation |
| | | 2. Capacity installed (6 + 3 water towers with solar panels operated by pumps + pipes) by the project (baseline = 0) |
| | | 3. Number of new acres of land irrigated (baseline = 12, target = 36) |
| R3 | In rural areas, 3000 students attending six schools will have adequate electricity provision (up to 2000 kWh per year) through solar energy to support light, IT and Internet teaching, and laboratory activities | 1. Schools with adequate annual electricity consumption level (6 × 2000 = 12000 kWh/year—baseline: none) |
| | | 2. Number of people and social activities benefiting from the electricity generated in the school centers |

In line with the IM&E framework previously described, we made two modifications of the original LFM: (i) project expected results are restructured to better visualize the *people-oriented* component, and (ii) *people-oriented* indicators are added at the level of results, specific goals, and overall objectives.

The new resulting matrix is clarified in Table 5.2. It is worthwhile mentioning that the review of the original LMF and the development of the new set of indicators have been based on the results of the baseline evaluation that we performed at the local level. Through the baseline evaluation, we identified the context in terms of basic information (initial situation, trends, organizational structures and processes already present, community's priorities, livelihood assets' availability, and access), energy-related information (initial energy condition of recipients and their energy needs), and new possible opportunities. By comparing the original set of indicators with the new one, it becomes clear how the IM&E allows monitoring of recipients' proactive roles within the project activities and assessment of their capabilities to correctly use and manage the new energy technologies.

**Table 5.2** New matrix resulting from the IM&E approach for the COOPI project

| | Project description | *People-oriented* indicators |
|---|---|---|
| Expected results | | |
| R1 | Component 1 (C1): | Output indicators R1_C1: |
| | –Three small and medium enterprises (SMEs) (20 persons involved) are available at local level and are able to provide ICSs to local households | –N° of privately purchased ICSs [N]<br>–N° of delivered ICSs by COOPI [N] |
| | Component 2 (C2): | Output indicators R1_C2: |
| | –1600 households have ICSs and are able to use them | –N° of ICSs damaged /not in use [N]<br>–N° of ICSs currently used [N] |
| R2 | Component 1 (C1): | Output indicators R2_C1: |
| | –Six farmer committees have the technical capability to irrigate 24 new acres with a PV plant<br>–Six farmer committees have the capability for economic and technical management of the irrigation system | –N° of irrigation systems (pump, water tower, irrigation pipes/channels) damaged /not in use<br>–Irrigated acres<br>–Annual expenditure for operation and maintenance (O&M) per plant [Malawi kwacha (MWK)/year plant] |
| | Component 2 (C2): | Output indicators R2_C2: |
| | –Six farmer committees have the technical capability to supply power for microbusinesses with a PV plant<br>–Six farmer committees have the capability for economic and technical management of the power plant | –N° of power systems (batteries, sockets) damaged /not in use<br>–N° of sockets in use (microbusinesses)<br>–Annual expenditure for O&M per plant [MWK/year plan] |
| R3 | Component 1 (C1): | Output indicators R3_C1: |
| | –Six schools have the technical capability to provide power to the provided IT facilities and lights with PV plants<br>–Six schools are able to use IT facilities (class, social activities) | –N° of power systems (batteries, sockets) damaged /not in use<br>—N° of IT facilities damaged /not in use |
| | Component 2 (C2): | Output indicators R3_C2: |
| | –Six schools have the technical capability to provide a power supply for microbusinesses with PV plants<br>–Six schools have the capability for economic and technical management of the power plant | –N° of sockets in use (microbusinesses)<br>–Annual expenditure for O&M per plant [MWK/year plant] |

## 5.4 Conclusions

Taking into consideration the recipients' roles within the project activities through *people-oriented* components, the proposed IM&E allows us to perform a more complete analysis of the project *results chain,* thus assessing the appropriateness of the implemented technologies. Two elements are key aspects of the new IM&E:

- Introducing three *people-oriented assumptions* (i.e., *participation, ownership,* and *empowerment*) allows us to highlight activities, strategies, and objectives adopted by recipients in response to project activities;
- Introducing two modifications to the standard LFM allows us to better visualize the *people-oriented components* and to add more *people-oriented indicators* at the level of results, specific goals, and overall objectives.

Monitoring and evaluating the *people-oriented* components is pivotal as they influence the success of energy projects and, thus, the implementation of energy technologies. Their analysis also allows us to develop new strategies that take into account the local context and the current situation of recipients, while also avoiding inappropriateness or a lack of involvement of recipients'.

## References

Adams, S., Annecke, W., Blaustein, E., Jobert, A., Proskurnya, E., Nappez, C. et al. (2006). A guide to monitoring and evaluation for energy projects (pp. 1–98). http://www.hedon.info/docs/MandEEDGuideFinalVersionEnglish.pdf. Accessed February 17, 2015.

Annecke, W. (2008). Monitoring and evaluation of energy for development: The good, the bad and the questionable in M&E practice. *Energy Policy, 36*(8), 2839–2845. doi:10.1016/j.enpol.2008.02.043.

Barkat, A. et al. (2002). *Economic and social impact evaluation study of the rural electrification program in Bangladesh* (p. 41). Dhaka, Bangladesh: Human Development Research Centre (HDRC). http://www.hdrc-bd.com/admin_panel/images/notice/1387698662.4a.%20%20economic%20and%20social%20impact%20evaluation%20study%20of%20rural%20electrification%20program%20in%20bangladesh.pdf. Accessed February 17, 2015.

Bensch, G., Kluve, J., & Peters, J. (2011). *Impacts of rural electrification in Rwanda* (p. 40). IZA Discussion Papers 6195, Institute for the Study of Labor (IZA).

Brent, A. C., & Rogers, D. E. (2010). Renewable rural electrification: Sustainability assessment of mini-hybrid off-grid technological systems in the African context. *Renewable Energy, 35*(1), 257–265. doi:10.1016/j.renene.2009.03.028.

Chambers, R., & Conway, G. (1992). Sustainable rural livelihoods: Practical concepts for the 21st century. *IDS Discussion Paper, 296*.

Cherni, J. A., Dyner, I., Henao, F., Jaramillo, P., Smith, R., & Font, R. O. (2007). Energy supply for sustainable rural livelihoods. A multi-criteria decision-support system. *Energy Policy, 35*(3), 1493–1504. doi:10.1016/j.enpol.2006.03.026.

Colombo, E., Masera, D., & Bologna, S. (2013). Renewable energies to promote local development. In E. Colombo, S. Bologna, & D. Masera (Eds.), *Renewable energy for unleashing sustainable development* (pp. 3–25). Springer International Publishing. doi: 10.1007/978-3-319-00284-2.

COOPI Cooperazione Internazionale. (2012). Energia. http://www.coopi.org/it/cosafacciamo/nelsuddelmondo/settori-d-intervento/energia/. Accessed February 17, 2015.

Danish Energy Management. (2009). Selected indicators for projects financed by the energy facility. http://www.energyfacilitymonitoring.eu/images/stories/publications/thematicfiches/tg_indicators_en.pdf. Accessed February 17, 2015.

Department for International Development—UK DfID. (1999). Sustainable livelihoods guidance sheets. http://www.eldis.org/vfile/upload/1/document/0901/section2.pdf. Accessed February 17, 2015.

European Commission. (2003). The EU energy initiative for poverty eradication and sustainable development (European Commission. Directorate-General for Development, Ed.). Office for Official Publications of the European Communities.
European Commission. (2004). Aid delivery methods—Project cycle management guidelines (Vol. 1). https://ec.europa.eu/europeaid/sites/devco/files/methodology-aid-delivery-methods-project-cycle-management-200403_en_2.pdf. Accessed February 17, 2015.
Fry Consultants Inc. (1970). Project evaluation and the project appraisal reporting system (p. 75). http://pdf.usaid.gov/pdf_docs/pnadw882.pdf. Accessed February 17, 2015.
International Energy Agency [IEA]. (2014). *World Energy Outlook 2014* (p. 748). OECD Publishing.
Khandker, S., B. Koolwal, G., & Samad, H. (2009). *Handbook on impact evaluation. Quantitative methods and practices* (p. 262). The World Bank. doi: 10.1596/978-0-8213-8028-4.
Kollmair, M., & Gamper, St. (2002). The sustainable livelihoods approach (p. 11). Input paper for the integrated training course of NCCR north-south. Development Study Group. University of Zurich.
Krantz, L. (2001). *The sustainable livelihood approach to poverty reduction. An introduction* (p. 44). http://www.sida.se/contentassets/bd474c210163447c9a7963d77c64148a/the-sustainable-livelihood-approach-to-poverty-reduction_2656.pdf. Accessed February 17, 2015.
Kremer, M., van Lieshout, P., & Went, R. (2009). *Doing good or doing better—Development policies in a globalizing world.* Scientific Council for Government Policy (Ed.). Amsterdam: Amsterdam University Press.
Kusek, J. Z., & Rist, R. C. (2004). *Ten steps to a results-based monitoring evaluation system* (p. 264). Washington DC: The World Bank.
McDonald, D. (1999). Developing guidelines to enhance the evaluation of overseas development projects. *Evaluation and Program Planning, 22*(2), 163–174. doi:10.1016/S0149-7189(99)00009-9.
Organisation for Economic Co-operation and Development [OECD]. (1991). Principles for evaluation of development assistance. DAC principles for evaluation of development assistance (p. 12). Paris.
Organisation for Economic Co-operation and Development [OECD]. (2005). *The Paris Declaration on Aid Effectiveness and the Accra Agenda for Action.*
Peters, J. (2009). Evaluating rural electrification projects. Methodological approaches. Ruhr Economic Papers 0136, Rheinisch-Westfälisches Institut für Wirtschaftsforschung, Ruhr-Universität Bochum, Universität Dortmund, Universität Duisburg-Essen.
Scoones, I. (1988). Sustainable rural livelihoods—A framework for analysis. *IDS Working Paper, 72.*

# Chapter 6
# Holistic Approach to Sufficient, Reliable, and Efficient Electricity Supply in Hospitals of Developing Countries: Cameroon Case Study

**Guy Merlin Ngounou, Michael Gonin, Nicolas Gachet and Nicolas Crettenand**

**Abstract** While health technology has shown constant improvements in industrialized countries, developing countries have not been able to take full advantage of this evolution, partly because of unstable power supplies. According to a World Health Organization study, grid failures are responsible for one-third of medical device breakdowns. Therefore, the global slogan "Health for All in the Third Millennium" requires a reliable and sustainable electricity supply in hospitals. This paper presents a power backup and electricity stabilization system that takes into account the technical constraints, as well as the socioeconomic factors, impacting the electricity supply in Cameroonian hospitals. The implementation of technological solutions has to be adapted to the socio-institutional context of the hospital. Preliminary sociological studies highlight the impact of organizational culture, hierarchy, and professional education on the way that technical equipment is installed and maintained, as well as the way that supply failures are addressed. From an economic perspective, technical weaknesses imply higher energy costs and lower revenues. Preliminary studies suggest that the costs incurred in the installation and maintenance of a stable electric system can partly be compensated through energy saving and additional medical treatments resulting from the increased availability of medical devices. The problem tree analysis method (PTAM) used in

G.M. Ngounou (✉)
Centre Universitaire de Recherche sur l'Energie pour la Santé (CURES), Ecole Nationale Supérieure Polytechnique de Yaoundé (ENSPY), Yaoundé, Cameroun
e-mail: ngounou.guy@energie-cures.org

M. Gonin · N. Gachet
Université de Lausanne, Lausanne, Switzerland
e-mail: michael.gonin@unil.ch

N. Crettenand
Ecole Polytechnique Fédérale de Lausanne (EPFL), Lausanne, Switzerland

N. Crettenand
Centre Universitaire de Recherche sur l'Energie pour la Santé (CURES), Ecole Polytechnique Fédérale de Lausanne (EPFL), Yaoundé, Cameroun

 
S. Hostettler et al. (eds.), *Sustainable Access to Energy in the Global South*,
DOI 10.1007/978-3-319-20209-9_6

this paper allows the identification of interactions between technical and socio-economic factors leading to electricity breakdown and, hence, to the development of more holistic solutions for the supply of electricity to hospitals. Because of its multi-dimensional nature, this project actively involves scholars from the North and South who are specialized in engineering, social and political sciences, and management.

## 6.1 Introduction and Purpose

While health technologies have shown constant improvements in industrialized countries, developing countries are not yet able to benefit fully from this evolution. Among others, unstable power supplies, inappropriate technologies, and lack of maintenance cause many electric medical devices to malfunction (see, e.g., Adair-Rohani et al. 2013). Regarding power supply in particular, regular blackouts cause frequent unavailability of medical devices and health services, while many devices are even damaged by electrical surges and transients. According to a vast study based on data collected from 33 hospitals in 10 developing countries (cf. Malkin 2007), on a total of 975 pieces of broken equipment, "*the most common single cause of failure was the power supply* (*29.9 %*)." Therefore, the global slogan "Health for All in the Third Millennium" involves a reliable and sustainable electricity supply.

Despite this observation, most projects that aim to enhance hospital infrastructure in developing countries have focused either on medical equipment or on providing some electricity-generating systems for those *not* connected to the national electricity network. In the same vein, most of the studies concerning electricity supply in sub-Saharan hospitals focus on the number of hours with access to energy and do not deal with electricity quality and reliability. This is problematic, as electricity supply failure impacts the sustainability and reliability of medical care and, hence, a population's health and well-being. This, in turn, indirectly impacts school attendance (and, hence, education level), as well as the national economy through the reduction of working population capacity.

In this paper, we consider the context in which electric medical equipment is used, that is, the technical and socioeconomic characteristics of the electricity supply within hospitals. We contend that a thorough understanding of this multi-dimensional context is essential to develop the power backup and electricity stabilization system necessary for high-quality medical care.

From a technical perspective, we discuss a hospital-specific network design, as well as the device-specific electricity need across time. From a socioeconomic perspective, we discuss the impact of organizational culture, hierarchy, and professional education on the way that technical equipment is installed and maintained as well as on the way that supply failures are addressed. Furthermore, we present preliminary reflections about solutions that would secure electric power in hospitals without necessarily increasing costs.

From an empirical perspective, our study focuses on district hospitals in Cameroon, a country that is often considered as "little Africa," suggesting that solutions developed there might work, with some adaptations, in many other developing and emerging countries, especially sub-Saharan countries.[1]

Section 6.2 presents the problem tree analysis method (PTAM) as a way to study complex issues and identify the core issue, its consequences, and its root causes. In Sect. 6.3, we discuss the technical and socioeconomic factors impacting electric power in hospitals and then integrate them into a problem tree that highlights their complex interconnections, pointing toward some specific root causes that need to be tackled. In Sect. 6.4, we sketch out potential solutions and discuss the limitations of the study, as well as further research.

## 6.2 Design and Methods

Because of the limited attention paid in the literature to the relationship between technical and engineering dimensions in electricity breakdowns on the one hand and socio-institutional, medical, and managerial ones on the other, a holistic approach of electricity failure requires (a) a thorough analysis of the existing technical situation in district hospitals, and (b) a participative approach involving local practitioners and scholars to better grasp the impact of the socio-institutional environment. Therefore, a multidisciplinary approach has been chosen to integrate insights from various disciplines in a single holistic framework, based on the PTAM.

### *6.2.1 The Problem Tree Analysis Method*

The PTAM is widely used by international development actors, and it represents an effective participatory tool to identify and structure problems faced by a community or an institution (e.g., Anyaegbunam et al. 2004; GTZ 1988, 1991). A main assumption underlying the PTAM is that the main issue is always interconnected with others issues. Through the PTAM, the causal links between these various issues are identified to (1) highlight the core problem, (2) define the root causes of this problem, and (3) understand the multiple consequences of the problem. The causes are identified by asking the simple question "Why does this problem occur?", while the effects stem from the question "What does the problem lead to?" As a result of this analysis, a kind of mind map, which looks like a tree, can be drawn. The main problem represents the trunk, the causes and their multiple ramifications are illustrated through the roots, and the consequences, with numerous

[1]See, e.g., http://www.rfi.fr/contenu/20091231-cameroun-une-afrique-miniature.

multiple ramifications, are found in the branches. This presentation highlights the root causes that need to be addressed to obtain a more profound impact. To take into account as many related issues as possible, the PTAM ideally requires a participative approach. Brainstorming and a focus group often allow for the identification of "hidden" issues and links.

### *6.2.2 Research Method*

Following the PTAM, we conducted a three-step study of the quality and reliability of the electricity supply in district hospitals in Cameroon. In the first step, information was gathered in regard to the technological and socio-institutional issues related to the electricity supply in district hospitals. A Cameroonian engineer visited various district hospitals to analyze the situation in regard to electric components, wiring, and management. The audit was conducted in four representative hospitals (public and confessional) located in different regions of Cameroon: a district hospital based in an urban area, two district hospitals based in a rural area, and a district hospital developing its activities with the help of international actors. These hospitals were selected because they all have sufficient electric equipment and electro-medical devices. They have between 70 and 120 beds, employ 40–90 people, and receive patients from all social classes.

The study used an audit protocol inspired from tools developed by the Powering Health initiative of the United States Agency for International Development (USAID) (2009). We interviewed key actors, such as mayors, directors, heads of service, doctors, nurses, and technicians. The questions asked referred to the exploitation and the management of the electric infrastructures, the equipment, the processes, the perceived quality of the electricity supply, and the impact of this quality. As no electrical diagrams were available, we also produced comprehensive electrical schemas for all hospitals. In addition, two Cameroonian experts—an engineer from an energy-efficiency service company and a sociology professor—provided important background information through their analysis of the socio-institutional and managerial factors impacting the electricity supply in district hospitals.

In the second step, meetings were organized with the research team and with external partners. In particular, a focus group in Yaoundé involved various persons representing the medical staff, the Ministry of Public Health, foreign companies that install and ensure the maintenance of medical equipment, and the hospital staff in charge of electric installation (Fig. 6.1). As a result, a problem tree that defined the main problem and proposed multiple root causes of the problem was drawn.

In the final stage, a tentative solution tree was developed that pointed to potential solutions for the main root causes of the problem. These solutions do not necessarily directly address the main problem (the trunk of the tree), but rather target root problems that will, in turn, impact the main issue.

**Fig. 6.1** Focus group meeting in Yaoundé. These meetings included entrepreneurs, engineers, doctors, and social scientists

## 6.3 Results

### *6.3.1 Characterization of Technical Problems*

#### 6.3.1.1 Quality of the Electrical Installation

The first important observation is that nobody seems interested in the electricity flow from the delivery point in the hospital to the use point. If a lamp can be lit, electricity is considered to be present—regardless of its quality. When an electric extension has to be made, no consideration is given to the maximum load capacity of existing cables. Consequently, 12 inappropriate extensions resulted in fires in the audited hospitals. Nineteen out of 41 buildings do not have earth connectors, 30 % of plugs are not sealed to the wall, and 17 cables lack insulation. In addition, the main electrical panel and the different departures of three audited hospitals contain intermingled cables, making it very difficult to look for the causes of an electrical problem. Ninety percent of the medical devices in the audited hospitals are connected to plugs by an intermediary of small ups or small voltage regulators, yet these devices are overloaded. They regularly break down and are not replaced until the financial service grants the necessary funds. Furthermore, the absence of an overall electric diagram of the installation does not allow for suitably balancing the loads between the phases in the case of an electric extension—sparks were observed on the principal departure of a phase in one of the audited hospitals because of an overload, while the other two phases were almost unused. Seventy percent of the

**Fig. 6.2** Marks of fire on an electric panel exposed to weathering

circuit breakers are exposed and can be switched off by anyone (Fig. 6.2). Finally, no protection against overvoltages was observed in these hospitals.

In addition to technical misconceptions, the technical staff seems to lack knowledge about the earthing system used. Safety devices, such as circuit breakers, are placed randomly, and the lack of spare parts regularly forces the staff to short-circuit damaged circuit breakers—seven such cases of short-circuiting circuit breakers were observed, two of which involved general circuit breakers. As a consequence of these technical misconceptions and maintenance problems, medical devices, as well as patients and staff, are insufficiently protected against overloads, short circuits, and insulation defects.

#### 6.3.1.2 Power Quality Issues and Consequences

The national Cameroonian electrical energy supply is insufficient and unreliable, with recurring power cuts ranging from 1 h to three days.[2] In all services audited, people complained about the frequent destruction of medical instruments because of bad electricity quality. Sixty percent of the interviewees affirmed that there is one power cut lasting from 10 min to 3 h every day. Furthermore, each recovery of the supply network is followed by an overvoltage that is responsible for 70 % of device

[2]http://isnblog.ethz.ch/development/cameroon-electric-dreams-for-development-by-2035.

**Fig. 6.3** Small uninterruptible power supplies are frequently used to provide emergency power to a load when the main electricity fails. Many of them are damaged by electrical surges

breakdowns. Medical equipment is protected against these surges through small uninterruptible power supplies (UPS) that ultimately break down after one year (Fig. 6.3).

An anecdotal, but dramatic, illustration of this recurring problem is the destruction of 50 % of the electric equipment of one hospital due to overvoltage coming from the grid in 2011. This is a problem commonly observed in sub-Saharan Africa, for example, in the Albert Schweitzer Hospital in Gabon (a country bordering Cameroon), where frequent breakdowns of radiology equipment are caused by the voltage drops and overvoltage of the public electricity network (Journal de l'association suisse Albert Schweitzer, Nouvelles de Lambaréné N°112, Janvier 2014). At peak hours, the equipment, such as sterilizers, remains unavailable due to the lack of energy, especially in rural areas, forcing the medical staff to use fire to sterilize the equipment. Two hospitals that have mortuary refrigerators also complained about the putrefaction of bodies resulting from longer power cuts —a situation that disappoints many bereaved families.

#### 6.3.1.3 Analysis of Energy Needs

Not only is the electrical energy supply insufficient and unreliable, it is also used without planning in the hospitals. The audit allowed for the establishment of a power budget for all devices, taking into account the coefficient of utilization and simultaneity. Among others, it is to be noted that in case of power cuts, only generators ensure electricity production, covering on average 69 % of daily energy needs of the four hospitals.

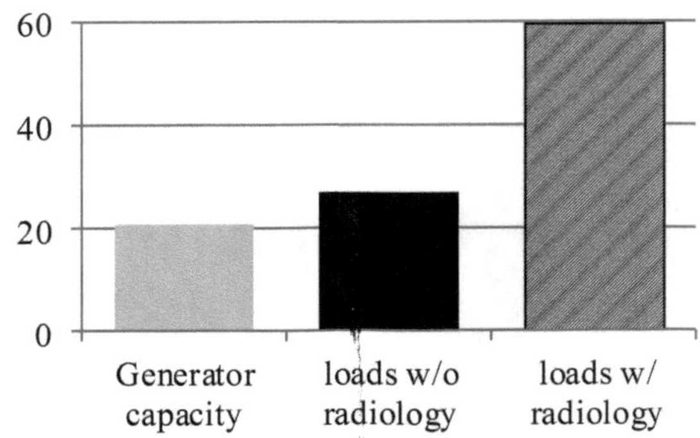


**Fig. 6.4** Average generator power and loads during power cut (kW)

Furthermore, the consumption load of radiology devices is largely above the capacity of the generator (Fig. 6.4). This overload causes additional voltage drops in the hospital and subjects the generators to hard working conditions that eventually lead to damage. This is the case in three out of the four audited hospitals. Furthermore, generators can function no more than an average of 10 h per day due to the lack of fuel.

### *6.3.2 Socio-institutional and Economic Factors Related to Hospital Electric Instability*

The shortcomings mentioned in Sect. 6.3.1, however, cannot be explained merely through technical factors. They also result from a lack of interest by the hospital directors for the management and maintenance of the electric infrastructure and equipment. There is no exploitation and/or maintenance plan and often one single (unqualified) person, without any formal authority, is in charge of all the maintenance within the hospital. Furthermore, there is a lack of financial resources and spare parts. For instance, in one of the audited hospitals, the two motors of a Caterpillar generator donated by the Swiss army were destroyed because the oil was not replaced.

In the same vein, the absence of incentives regarding energy use can lead to an inefficient use of that energy. For instance, as the national grid electricity bill is paid directly by the Ministry of Public Health, there is no incentive to switch lights off during the day. As a result, external lights are systematically lit during the day in all the audited hospitals. However, in the eventuality of power cuts, the hospitals have to buy fuel for the generator using funds from their own budget. Unfortunately, the only functional generator in one audited hospital automatically starts during peak hours due to the voltage drops of the grid. This situation requires the hospital to use their revenues to buy more fuel for the generator even though energy from the grid is available. The consequence of this additional expenditure is a reduction in the production bonuses for the medical staff, a situation that frustrates doctors. The data collected during the audit allowed us to sketch the distribution of the annual energy budget, with the biggest part being paid directly by the Ministry (Fig. 6.5).

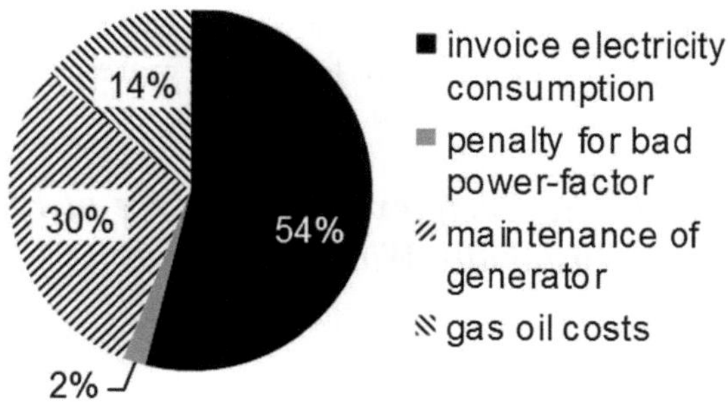


**Fig. 6.5** Distribution of the annual energy budget [27,000 Swiss francs (CHF)]

A conclusion of a study made by the "Institut de l'énergie et de l'environnement de la Francophonie" (2006) on health facilities in developing countries is that the issue of energy savings is more important in health institutions in developing countries than in industrialized countries. The study shows that the energy consumption of major national and regional hospitals (fuel and electricity) in developing countries is estimated to constitute about 5 % of the overall budget, although it can reach 10 %.

By implementing adequate measures to monitor energy use, coupled with some changes in equipment and process management, not only could medical equipment availability be improved but energy use could also be drastically reduced, leading to major savings for hospitals. In regard to the socio-institutional factors, multiple causes have been identified that contribute to the current deficiencies, but they also show potential for developing novel solutions.

However, such changes are often difficult to implement because they go against the established organizational structure and culture. As mentioned, technical issues are often considered to be secondary by the top management, and (unqualified) technical staff is minimized. Furthermore, the technicians often seem disconnected from the rest of the hospital team, including the administrative staff. Therefore, their sphere of influence is highly limited and they lack the capacity to make appropriate decisions and to implement new routines and processes regarding the use of the electric installation. Furthermore, the absence of an incentive scheme and of clear responsibility regarding energy use and installation maintenance is also valid for the entire administrative and medical staff. This leads to situations in which no organizational member feels responsible for the problem and, if one does, this person often lacks the capacity and authority to tackle the issue effectively. More broadly, the regional culture encourages people to avoid intervening in issues in which they do not feel fully confident, in order to avoid being held responsible for the problem. Finally, fatalism prevents proactive "insurrection" against specific dramatic problems, a rage that could sometimes provide motivation to find new and innovative solutions.

In the broader Cameroonian context, preliminary studies suggest that the government is willing to support small- and medium-sized enterprises (SMEs) and new initiatives—a factor that might be relevant to the potential solutions that will be discussed in Sect. 6.4.1.2. It is also noteworthy that patients often pay for additional services in hospitals—a point that might open an avenue for additional revenues in relation to additional comfort-related electricity consumption (e.g., the use of light for comfort or the use of televisions in the patient rooms).

## 6.3.3 The Complex Problem Tree of Electricity Supply Instability Within Hospitals

### 6.3.3.1 Causes

Figure 6.6 presents the problem tree resulting from the studies, brainstorming, and focus group. It shows that the core problem is the non-satisfactory coverage of electric power needs. While the different boxes of the figure have been addressed in the previous sections, a few important points regarding their interconnections deserve attention at this stage. First, there are some specific technological problems that need to be addressed regarding the conception and installation of electric equipment to ensure that the electricity supply is sufficient in terms of quantity and quality.

However, the second observation is that there are various socio-institutional and managerial factors that lead to an unsatisfactory energy supply and that interact with the technical dimensions. Among others, the absence of strategic planning regarding the importance of the various devices and the possibility to connect them simultaneously to the network compared to the wire and generator capacity seem to play an important role. In the same vein, the absence of clear delegation of responsibility (and the corresponding capacity and authority to act) leads to an inefficient use of energy and to the absence of maintenance planning. In most structures, no plan, routine, or roadmap exists that would guide the medical and administrative staff in the proper manner to use, connect, and switch on/off specific electric devices, especially in times of shortage.

At a more general level, education and vocational training programs for the staff to acquire the required competencies might be lacking. In addition, the prevailing donation logic within many public or private international organizations tends to prevent a systematic reflection about the total electric capacity of the hospital, as well as about the development of incentives regarding the careful and parsimonious use of the devices. This contributes to the overexploitation of the electric capacity, as well as to poor operational management, which regularly leads to the breakdown of the sponsored devices.

### 6.3.3.2 Consequences

The main problem regarding the electricity supply in Cameroonian district hospitals, identified as non-satisfactory coverage of electric power needs, not only has multiple causes, but it also impacts, in various ways, the processes and performance of the hospitals. For instance, many medical devices (e.g., radiography equipment) can be damaged by electrical instability. Moreover, the lack of light in surgical

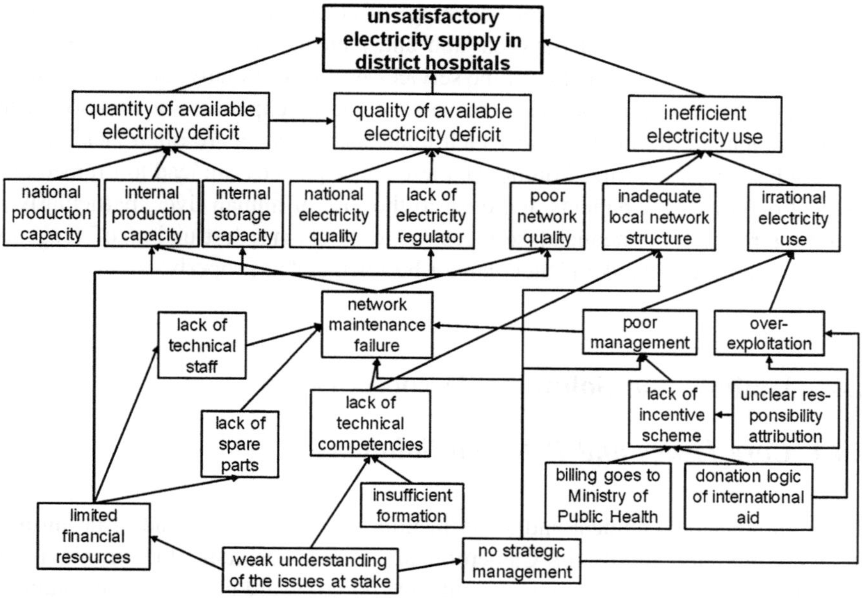


Fig. 6.6 Problem tree of the energy supply in Cameroonian district hospitals: lower part (causes)

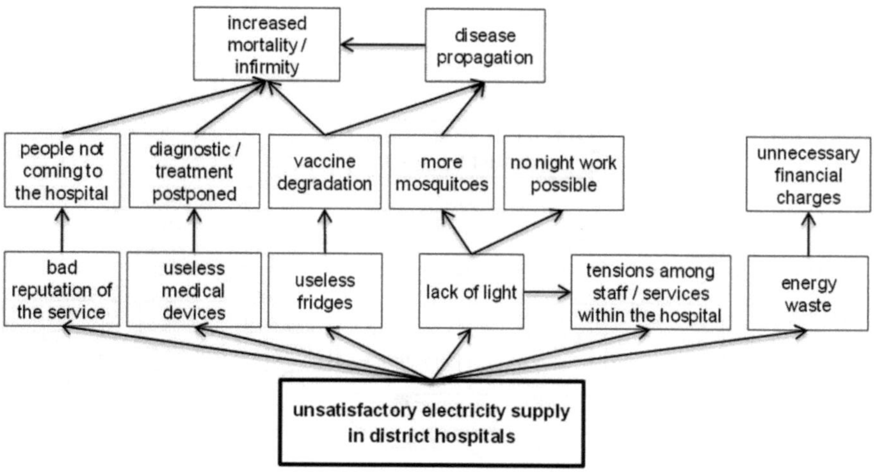


Fig. 6.7 Problem tree of the energy supply in Cameroonian district hospitals: upper part (consequences)

theaters and patients' rooms makes night work impossible and increases the number of mosquitoes and, hence, the risk of malaria. All these problems, in turn, decrease the quality and reliability of the services provided by hospitals. They might also discourage people from seeking medical help before their injury or illness becomes too severe to be treated (Fig. 6.7).

In addition, the non-satisfactory coverage of electric power needs might lead to tensions among the staff or between staff and patients, resulting in reduced motivation and, hence, lower quality of the services provided. Finally, the inefficient use of energy leads to additional and unnecessary energy expenses—a situation especially problematic in a context in which financial resources are often lacking even for very basic medical care. Combined, these consequences can lead to irremediable medical outcomes for patients and even to the loss of human lives, particularly in operating rooms and neonatal services, as well as the destruction of reagents, vaccines, and blood banks if they cannot be refrigerated properly.

## 6.4 Analysis and Solutions Perspectives

### *6.4.1 Core Issues and Potential Solutions*

By combining the technical and socio-institutional factors that lead to an unsatisfactory electricity supply in Cameroonian district hospitals, the problem tree developed in Sect. 6.3 offers new insights that point toward new potential solutions. Among others, it highlights that most of the roots also contain factors that relate to managerial and socio-institutional dimensions. Consequently, the solution tree must include both technical and socio-institutional dimensions (see Fig. 6.8).

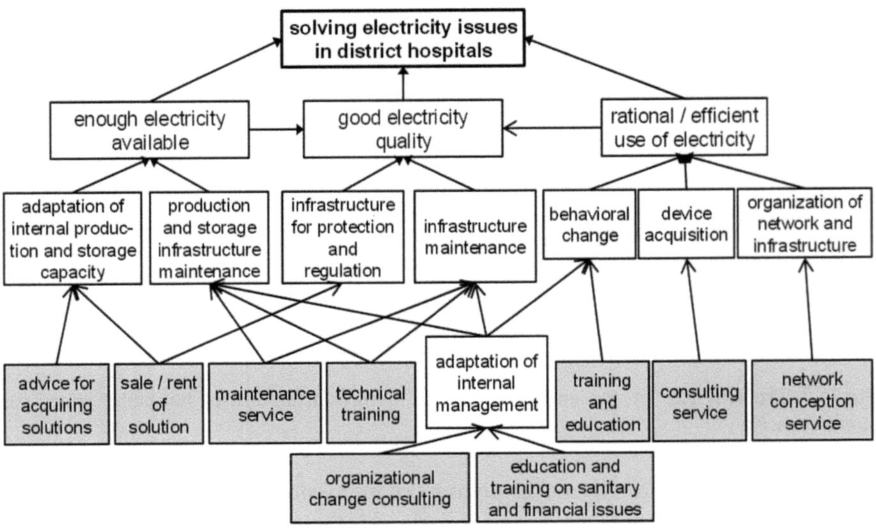


**Fig. 6.8** Preliminary solution tree

#### 6.4.1.1 Technical Solution for Adequate Power Backup and Electricity Stabilization

The audit enabled us first to note that the energy supplied at off-peak hours could be stored in accumulator batteries and reused at peak hours when the energy provided by the grid becomes insufficient. Second, in the case of a power cut, all the hospital loads are supplied by a generator that cover an average of 69 % of daily energy needs for the hospital while being overloaded and at the risk of failure. The analysis also revealed that the energy available could be sufficient to ensure the provision of vital services during an entire day if needed—with no overload and risk of failure—in the case of wise energy use. An analysis of the various electricity needs worked out from a medical perspective developed in collaboration with all heads of services (26 Cameroonian doctors) and Dr. Beat Stoll[3] enabled the identification of three different types of loads:

- *Vital load.* The load in which any interruption can endanger a patient's life (e.g., a respirator);
- *Essential load.* The load in which interruptions make a diagnosis extremely difficult (e.g., in the radiology unit);
- *Useful load.* The non-necessary load in the case of insufficient power (e.g., room ventilation during a hot period).

It is noteworthy that the prioritization of loads must take into account many local sociocultural aspects. For example, a lamp in a room is perceived differently in an urban zone than in certain rural zones. In one of the rural hospitals audited, some nurses and doctors do not administer care in rooms that have no light. The reason lies in stories of sorcery and malefic spirits that operate in the darkness and which would have been at the origin of certain patient deaths.

After having hierarchically arranged the hospital loads, we computed the daily average distribution of energy needs (Table 6.1, *column 1*) according to each type of load (*column 3*) and the corresponding required power at peak hours (*column 2*). Based on this, we propose to segment loads by linking the various medical needs to specific medical device technologies (*column 4*). We first distinguish between the useful loads, which do not need secured power, and the vital and essential loads that need secured power. Secured power is further subdivided into two subgroups: secured power without interruption (vital and essential loads supplied without interruptions) and secured power that tolerates short interruptions while waiting for the generator to be started (vital and essential loads that tolerate short interruptions). The latter category includes, for instance, medication fridges that are vital, yet have a thermic inertia and can, hence, withstand short power cuts.

Considering the daily average need and the average peak power need for each type of load, the power of the generator is sufficient to supply a vital and essential

---

[3]Dr. Beat Stoll worked for many years in Cameroonian hospitals and is, among others, the Chief Medical Officer of the EssentialMed Foundation (http://www.essentialmed.org).

**Table 6.1** Load segmentation according to the types of medical needs and the types of technologies

| Segmentation based on the type of medical needs | | | Segmentation based on the type of medical device technologies | |
|---|---|---|---|---|
| *Column 1* | *Column 2* | *Column 3* | *Column 4* | *Column 5* |
| Average daily needs (%) | Peak power (kW) | Type of load | Peak power (kW) | Group of segmentation |
| 35 | 7.5 | Vital | 4 | **Secured**, *no interruption* |
| | | | 3.5 | **Secured**, small interruption tolerable |
| 47 | 12 | Essential | 1 | **Secured**, *no interruption* |
| | | | 11 | **Secured**, small interruption tolerable |
| 18 | 6 | Useful | 6 | **Non-secured** power |

load during an entire day at no overload and risk of failure. However, this excludes the X-ray machine that must have its own power source to avoid overloading the generator. From a technical perspective, we obtained a refined segmentation by conciliating the medical needs according to the load types (*columns 4 and 5*).

The advantage of this solution is that, in the case of a crucial deficit of energy, for example, a lack of fuel, the useful and/or the essential loads can be temporarily switched off. Likewise, since the size of accumulator batteries can be reduced if they supply only the devices that do not tolerate interruptions, our solution foresees to separate the load supply with interruptions from loads that tolerate short interruptions. Furthermore, the solution is designed to protect the secured loads from grid disturbances (overvoltage, voltage drops, and harmonics), to compensate for the short interruption on secured loads without interruptions until the generator is started, and to contribute to reducing fuel consumption of the generator by controlling the automatic starting. The solution (Fig. 6.9) is based on the static technology of UPS supplies and more precisely on the online topology,[4] which has the advantages of being ideal for equipment that is very sensitive to power fluctuations, as well as being redundant.

A frequent limitation of the durability of traditional power backup systems is the use of lead acid batteries that have a reduced life span in hot environments and/or when used with deep cycles. Therefore, we have chosen the particular technology of nickel–metal hydride batteries (NiMH),[5] which offer a longer life span than lead batteries in situations of high discharge, are resistant to temperatures up to 60 °C, and do not produce explosive gas.

The suggested solution includes protection against voltage surges and has an estimated life span of at least 10 years. The capital costs for all the system

[4]See, e.g., http://en.wikipedia.org/wiki/Uninterruptible_power_supply for an introduction.

[5]The specific solution was developed based on work by the "Distributed Electrical System Laboratory" (see http://desl-pwrs.epfl.ch).

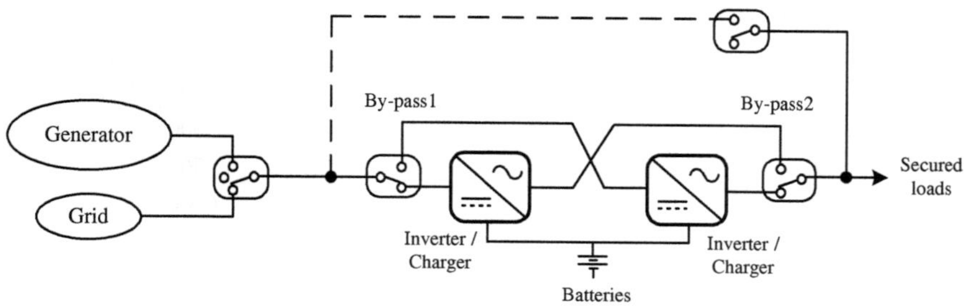


**Fig. 6.9** Diagram of the proposed power backup and electricity stabilization

components are based on public price quotations obtained from manufacturers, including the typical discount of integrators. These costs are likely to change for the actual system quotes due to many market factors. Therefore, the costs used in the analysis are indicative. The estimated cost of installation is based on the number of hours of work and hourly cost of a technician in Cameroon. The preventive maintenance of the system can be ensured by a technician in the hospital (after training) with a monthly maintenance budget. The corrective maintenance cost over 10 years should cover the damaged units. Indeed, the two main elements of the system are the power converters and the batteries. The NiMH batteries that we have selected to realize this system require no maintenance, are more resistant to overload (ARE 2013), and have an estimated life span of more than 10 years. Only one of the two power converters is exposed to the disturbances of the network. If we assume that we must replace a damaged power converter during its life span, then the replacement cost is estimated as the price of the device and the commission cost is calculated according to the number of intervention hours.

The number and characteristics of the system components are sized according to the total power of the essential equipment that has to be protected. The total power of equipment to be protected depends on each audited hospital and is given in Table 6.2. Table 6.3 details the total cost estimate for 5 kW and ten years exploitation.

Compared with the average annual energy bills of the audited hospitals, the suggested solution represents 14 % of these costs, including operation and maintenance costs. Indeed, the average annual electricity bill of the hospitals is about 15,000 CHF, and the average annual costs for fuel and maintenance for the

**Table 6.2** Total power of the equipment to be protected in each hospital

| Hospital | Total power equipment to be protected (kW) | Total cost over 10 years (investment + exploitation) |
|---|---|---|
| District hospital in an urban area | 3 | CHF 27,500 |
| District hospital in an urban area with the help of international actors | 4 | CHF 37,000 |
| Confessional hospital | 5 | CHF 44,700 |
| District hospital in an urban area | 5 | CHF 44,700 |

**Table 6.3** Indicative budget for a typical installation of 5 kW and exploitation over 10 years

| Parameters | Hours | Labor cost (CHF/h) | Monthly cost (CHF) | Total cost Over 10 years (CHF) |
|---|---|---|---|---|
| Capital costs | /// | /// | /// | 32,000 |
| Installation costs | 70[a] | 8 | /// | 560 |
| Preventive maintenance costs | /// | /// | 50 | 6000 |
| Corrective maintenance costs | | | | |
| Estimated replacement costs | /// | /// | /// | 6000 |
| Costs of commissions | 14 | 10 | /// | 140 |
| Total | | | | 44,700 |

[a]10 days × 7 h

generators are about 12,000 CHF. The expenses of installing the suggested solution could be covered by cost savings from energy-efficiency and savings measures.

However, while drastically increasing the quality and reliability of the power supply, the proposed technical solution also implies changes regarding the socio-institutional aspects of power management, as discussed in the following section.

#### 6.4.1.2 Elements of Socio-Managerial Solutions for Adequate Power Backup and Electricity Stabilization

From a socioeconomic perspective, the study points to the importance of adapting specific patterns, job descriptions, management strategies, as well as education and vocational training, to ensure that all the roots of the problem are addressed in concerted ways and to prevent the failure of the implementation of the technical solution due to the non-adaptation of the organizational structures. For instance, the repartition of the devices into three categories, useful, essential, and vital, requires a strategic analysis of the needs and functions of the services, an experienced engineer to design the corresponding electric network, and a skillful deployment of the wires. This, in turn, might require important investments in the education and training of technical as well as managerial and medical staff to evaluate correctly the various needs in the case of an energy breakdown.

Managerial skills are also needed to plan maintenance, fuel stocks, and device replacement. More generally, changes in incentive schemes can contribute to ensuring that all actors contribute to the quality, safety, and reliability of the electric network, both under normal and grid failure conditions. Finally, awareness training might help actors understand the importance of reliable energy and, consequently, of its parsimonious and well-planned use to reduce costs and increase stability. This, in turn, will impact the quality and reliability of the healthcare services provided.

These changes in institutional and organizational culture are important in regard to the financial feasibility of the deployment of the solution. The proposed solution, as well as its maintenance, costs many tens of thousands of Swiss francs. However, by implementing the structural and cultural changes, much of this money can be recuperated through energy cost savings of up to 20 %, as well as through increased "productivity" within the hospitals. In addition, as the solution prevents medical devices from being damaged, the cost of replacement of the medical devices can be drastically reduced. Finally, the solution ensures the continuity of medical care even during grid failure, thereby allowing for additional patients to be treated and for increased staff motivation. These factors, in turn, might lead to increased reputation and additional patients.

From an entrepreneurial perspective, the additional income, combined with adapted training, leads to new opportunities to create service and consulting firms. These might offer services such as the conception, installation, and maintenance of the electrical network, as well as power management and organization change.

From a broader perspective, reflections are needed on the conditions set by international organizations that donate medical devices to hospitals. A more thorough verification of hospital electricity capacity, but also of security devices to protect medical equipment from bad electricity quality—and, if needed, the donation of adapted devices to protect medical equipment—might contribute to increasing the life span of these devices, thereby drastically increasing their impact on the quality of hospital medical care.

### 6.4.2 Limitations and Further Studies

This study represents a preliminary work focused on four district hospitals and involving a limited number of stakeholders. Additional hospital audits and the inclusion of other stakeholders in the discussion of the problem tree and the solution tree might increase the reliability of the results and of the links identified. Measurement campaigns are also needed to gather data on the power supply (power availability and power quality) and electricity need (power consumption dynamic profiles).

Furthermore, while the technical solution sketched out in this article is undergoing preliminary tests, the managerial and socio-institutional aspects need further study regarding the ways the proposed changes can be implemented so that they impact the quality and reliability of the healthcare services in the long run.

Despite the need for additional confirmatory studies, this article contributes to clarifying the relationship between technical development and managerial challenges in the deployment of solutions for the stabilization of the energy supply in hospitals. As such, it points to innovative solutions to tackle unsatisfactory energy issues in district hospitals connected to the national grid by relating technical issues to specific socio-institutional and managerial aspects of hospital and energy management. It also challenges international aid agencies and nongovernmental organizations to ensure that the hospitals receiving medical devices through their agencies have the technical and managerial capacities to ensure the provision of proper electric resources and the maintenance of the devices.

## 6.5 Conclusion

In Cameroon, the technical specifications of equipment in district hospitals take into account the evolution of technology in the field of hospital equipment and undergo periodic updates (Ministère de la santé public du Cameroun 2004). Taking into account the vulnerability of these medical devices faced to grid disturbances and the unavailability of vital medical devices due to repetitive blackouts and the automatically starting of generators in case of grid voltage drops, the investment in an efficient electricity supply in district hospitals in Cameroon will reduce the unavailability of medical devices by at least 30 %. It will compensate for the short interruption in protected equipment until the generator is started, and it will also contribute to reducing the maintenance costs and fuel consumption of generators. All this will lead to increases in production bonuses for the medical staff, and it will improve the quality of care. In contrast to traditional power backup systems that use lead acid batteries, the developed solution uses NiMH batteries, which have the advantage of a longer life span in hot and wet climates for applications such as the supply of emergency power in hospitals, which requires many deep cycles. The solution has been tested in the laboratory and is being tested in the field. Considering the results obtained, the solution could significantly improve the quality of the power supply in hospitals. This would improve the quality of care and confidence of patients who often renounce treatment in hospitals because of a lack of functioning equipment.

Compared with the average annual energy bills of the audited hospitals, the suggested solution represents 14 % of the electricity, fuel, and maintenance costs. The suggested solution could be financed through cost savings thanks to adequate energy-efficiency and savings measures. However, it should be noted that after installing such a system in a district hospital, some changes in institutional and organizational culture are important for proper operation of such an electrical system.

**Acknowledgments** This work was conducted by the Université de Lausanne (UNIL), Ecole Polytechnique Fédérale de Lausanne (EPFL), and Ecole Nationale Supérieure Polytechnique de Yaoundé (ENSPY). The authors are grateful for the facilities offered to them in these universities. We thank the Cooperation & Development Centre (CODEV) and the Network of Excellence in Engineering Sciences for the French-speaking Community (RESCIF) for their ideas and for having agreed to supervise the research tasks regarding the reinforcing of primary healthcare systems in Southern countries with a reliable and sustainable electricity supply and with appropriate equipment. Part of the work was further realized thanks to the funding granted by the joint EPFL-UNIL CROSS program. Finally, the authors would like to thank the anonymous reviewers and the editor for their very constructive remarks that allowed us to improve the quality of the final manuscript.

## References


Anyaegbunam, C., Mefalopulos, P., & Moetsabi, T. (2004). *Participatory rural communication appraisal. A handbook* (second Ed.). SADC Centre of Communication for Development Harare and Food and Agriculture Organization of the United Nations (FAO), Rome.

Adair-Rohani, H., Zukor, K., Bonjour, S., Wilburn, S., Kuesel, A. C., Hebert, R., & Fletcher, E. R. (2013). Limited electricity access in health facilities of sub-Saharan Africa: A systematic review of data on electricity access, in sources, and reliability. *Global Health: Science and Practice, 1*(2), 249–261.

Alliance for Rural Electrification [ARE] (2013). Energy Storage Campaign.

Gesellschaft für Technische Zusammenarbeit [GTZ] (1991). *Methods and instruments for project planning and implementation*. Eschborn: GTZ.

Gesellschaft für Technische Zusammenarbeit [GTZ] (1988). *ZOPP (An introduction to the method*). Eschborn: GTZ.

Institut de l'énergie et de l'environnement de la Francophonie (2006). La maîtrise de l'énergie dans les établissements de santé. *Fiche technique Prisme no. 4.* http://www.ifdd.francophonie.org/docs/prisme/Fi-ME%20en%20ESante.pdf. Accessed 5 February 2014.

Journal de l'association Suisse Albert Schweitzer (2014). Nouvelles de Lambaréné, 112(Janvier).

Malkin, R. A. (2007). Design of health care technologies for the developing world. *Annual Review of Biomedical Engineering, 9*, 567–587.

Ministère de la santé public du Cameroun (2004). Propositions de listes standard et de spécifications techniques d'équipements des formations sanitaires de 6ème, 5ème et 4ème au Cameroun.

United States Agency for International Development [USAID] (2009). A guide to the energy audit for specifying energy systems in off-grid health facilities. USAID Tools. http://www.poweringhealth.org/index.php/resources/tools/energy-audit-spreadsheet. Accessed 28 January 2014.

# Part III
# Up-Scaling Energy Solutions

# Chapter 7
# Scaling-Up Sustainable Pro-poor Energy Solutions: Addressing Stumbling Blocks

**Albrecht Ehrensperger and Susanne Wymann von Dach**

**Abstract** Affordable, reliable, and modern pro-poor energy solutions play a key role in achieving sustainable development in emerging economies. While such solutions have been successfully tested in numerous protected niches and "proto-markets," their diffusion beyond these niches has remained very limited. There are a number of factors explaining the widespread failure to change prevailing energy supply regimes. An actor-oriented perspective (Wiesmann et al. in Research for sustainable development: foundations, experiences, and perspectives: Perspectives of the Swiss National Centre of Competence in Research (NCCR) North-South. University of Bern, Bern, pp. 231–256, 2011) helps demonstrating that these factors are primarily linked (1) to action itself, viewed as the interplay between actors' existing practices, their means, and assets, and the meaning actors attribute to action; and (2) to external dynamic social, economic, political, and ecological conditions of action influencing actors' perceptions. The four contributions of Part III (Chaps. 8–11) identify the stumbling blocks hindering the successful scaling-up of pro-poor energy solutions. The solutions proposed by the authors focus on the dynamic conditions of action (Kung et al., Chap. 8; Zalengera et al., Chap. 9), on local actors' perceptions, means, and activities (Mirza, Chap. 10), and on the utilization of a comprehensive planning and decision-making framework to improve the sustainability of pro-poor energy solutions (Jain and Kattuman, Chap. 11).

A. Ehrensperger (✉) · S.W. von Dach
Centre for Development and Environment (CDE), University of Bern, Bern, Switzerland
e-mail: albrecht.ehrensperger@cde.unibe.ch

S.W. von Dach
e-mail: susanne.wymann@cde.unibe.ch

 
S. Hostettler et al. (eds.), *Sustainable Access to Energy in the Global South*,
DOI 10.1007/978-3-319-20209-9_7

## 7.1 Access to Energy: A Key Issue of Sustainable Development

Access to energy is increasingly recognized as a major prerequisite to overcoming persistent poverty in developing countries, especially in rural areas (Nussbaumer et al. 2012; Secretary-General's High-Level Group on Sustainable Energy for All 2012). The United Nations Millennium Project (Modi et al. 2006) provided ample evidence that access to improved energy for household needs is linked to progress toward the Millennium Development Goals. More recently, the Working Group on Sustainable Development Goals has proposed a list of goals among which Goal 7 emphasizes the role of affordable, reliable, and modern energy in achieving sustainable development (Open Working Group on Sustainable Development Goals 2014). And development practitioners have found that poor people's access to energy at the household level can help them improve their income and is indispensable for the provision of some community services (Practical Action 2013).

Consequently, efforts have been made in the last decades to develop improved energy technologies such as efficient stoves, fireless cookers, fuel briquettes made from biomass residues, solar cookers, domestic biogas systems, solar photovoltaic systems, village-level off-grid rural electrification systems, and more. Most of these initiatives not only aim to improve local communities' access to reliable renewable energy, but also work toward broader sustainability objectives, including alleviation of widespread poverty, reduction of health hazards from indoor air pollution, and mitigation of environmental degradation caused by overuse of woody biomass and other natural resources to cover energy needs.

## 7.2 Long-Term Viability and Successful Scaling-Up of Improved Energy Technologies

Improved energy technologies have been implemented successfully in niches, under specific favorable conditions and with external support. But their diffusion beyond these niches has remained limited, and prevailing energy supply regimes have hardly changed—irrespective of whether they are traditional and locally structured or built up and administered by national states or large companies. In fact, a good number of improved energy initiatives implemented in protected niches or "proto-markets" (Schot and Geels 2008) in developing countries actually failed the test of time and were shut down shortly after external financial and logistical support was discontinued. These failures raise the following questions: What tangible benefits can local communities derive from improved energy solutions in the long term? And how can promoters ensure long-term functioning of the systems they introduce?

The poor viability of improved energy projects and their inability to spread beyond protected niches seems to be caused by a number of factors. Romijn et al. (2010)

analyzed failed biomass energy projects in rural India using a learning-based framework that combines components of earlier work by Korten (1980), Uphoff et al. (1998), Douthwaite (2002). The framework defines best practices for the development projects with regard to project design and management, project implementation culture, stakeholder participation, and consideration of external factors. The authors found that the observed energy projects in rural India deviated substantially from these best practices. A crucial aspect was the lack of efforts to foster communities' self-reliance. In several cases, implementing agencies had taken technology as a starting point of reflection and investment without clarifying local people's actual needs; this resulted in low ownership among local actors. Finally, implementers had frequently underestimated the hampering effects of local rivalries between potential beneficiaries, as well as the expansion of government-subsidized rural electrification and its stifling effect on actors' willingness to invest in local solutions. Mogaka et al. (2014) found that in some cases, especially when projects are steered by nongovernmental development organizations, technical solutions are promoted and adopted by beneficiaries in the absence of an established and reliable market for these solutions. One such example is the introduction of *jatropha* for biofuel in parts of East Africa. Large-scale investors and smallholder farmers who had started producing *jatropha* around 2005 abandoned the energy crop a few years later due to insufficient returns and the inability to compete with fossil fuels.

## 7.3 An Actor-Oriented Perspective for Addressing Stumbling Blocks

The human actor model (Fig. 7.1) proposed by Wiesmann et al. (2011) differentiates between the dynamic conditions of action and action itself, which is viewed as an interplay between actors' practices, the means and assets available to them, and the meaning they attribute to certain actions. From this actor-oriented perspective, decisions to engage in an improved energy project or to maintain an existing village-level energy solution are taken against the background of dynamic social, economic, political, and ecological conditions that are part of a wider energy production and consumption regime (in the sense of Schot and Geels 2008). In the examples provided in Sect. 7.2, the Indian government's policy to expand the national grid to remote rural areas or the lacking market for *jatropha* biofuel in East Africa are dynamic political and economic conditions that directly impact local actors' decisions to invest—or not to invest—in village-level improved energy technologies. At the same time, actors' decisions are co-determined by their own assets, means, current activities, and practices. When promoters and implementers of pro-poor energy solutions focus primarily on technology while neglecting local energy needs—as described by Romijn et al. (2010; see Sect. 7.2)—they tend to overlook the multifaceted rationales underpinning local actors' decisions and actions.

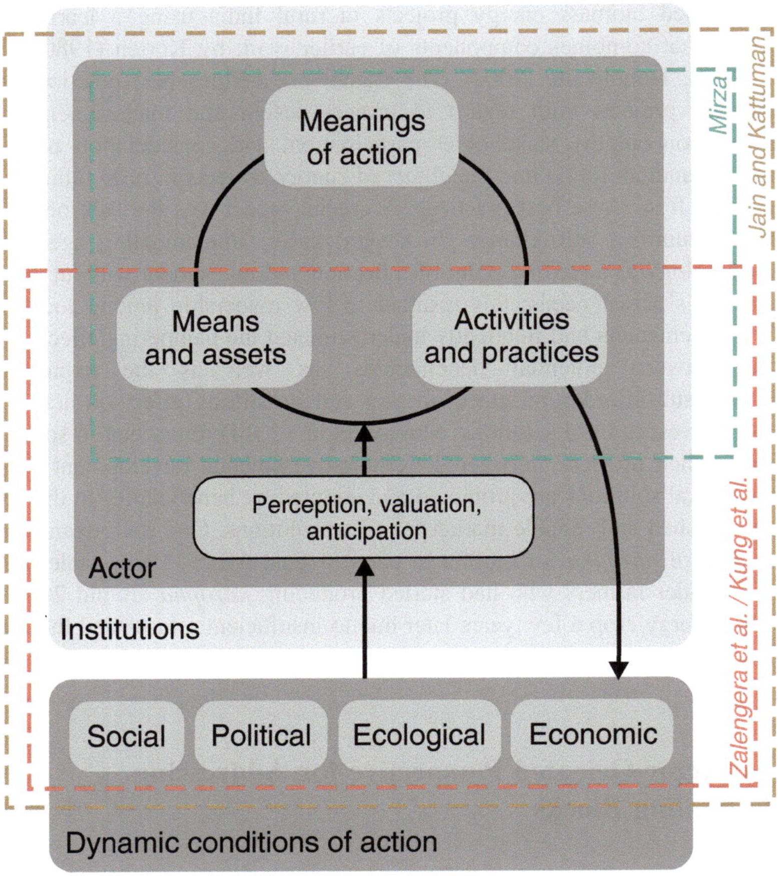


**Fig. 7.1** Model of human action and of its determinants. The four chapters of Part III contribute to a better understanding of the dynamic conditions of action and of the interplay between action, means, and meanings of action (Adapted from Wiesmann et al. 2011)

And when they fail to build self-reliance and a sense of ownership among local communities, they have not adequately factored in these communities' means and assets for maintaining the given energy solution.

## 7.4 The Contributions of Part III

The next four chapters investigate stumbling blocks that hinder the successful scaling-up of improved energy technologies. They do so from different perspectives. The remainder of this introductory chapter uses the actor model by Wiesmann et al. (2011) to bring together the different insights and highlight each chapter's specific contributions to a better understanding of the multifaceted adoption and diffusion processes related to sustainable energy. As shown in Fig. 7.1, Chaps. 8 and 9 focus on dynamic economic, social, and political conditions that co-determine local stakeholders' actions; Chap. 10 looks at local actors' perceptions, means, and activities to assess reasons for project failures; and Chap. 11 proposes a comprehensive planning and decision-making framework that touches on all aspects of the actor model by identifying determinants of success on the background of key principles of project implementation and important scaling factors.

*Kevin S. Kung, Samuel Wanderi Rigu, Steve Kariithi Karau, Kamau Gachigi, and Libby McDonald* (Chap. 8) propose a micro-enterprise model for the production of renewable briquettes from agricultural residues in Kenya. They advocate decentralized household-level char production using small, inexpensive kilns, coupled with more centralized and professionalized briquetting infrastructures to ensure efficient production and quality output. The authors base their suggestion on an analysis of socioeconomic factors informing the choices of potential consumers, such as price and fuel quality.

*Collen Zalengera, Richard E. Blanchard, and Philip C. Eames* (Chap. 9) stress the importance of policy intervention and economic development as dynamic factors for improving rural access to electricity. Their research shows that current grid electricity is too expensive for households on Likoma Island in Malawi and that proposed low-price alternatives are often poorly adapted to local energy needs. They argue that household-level solar photovoltaic systems could be a viable solution—provided that households' purchasing power increases and that the government extends eligibility for feed-in tariffs to small-scale systems.

*Bilal Mirza* (Chap. 10) analyzes local actors' needs and perceptions to explain rural communities' high expectations of village-level solar systems and the poor sustainability of projects implementing such systems in Pakistan. He identifies the technology-centered approach of project promoters, the lack of consideration for people's actual energy needs, and the lack of investment in maintenance and follow-up support as important reasons for project failure, high levels of resentment, and low ownership among adopter communities.

*Abhishek Jain and Paul Kattuman* (Chap. 11) address the question of sustainability and scaling-up of pro-poor renewable energy systems from a systemic perspective, considering beneficiaries' perceptions, rationales of action, and means, as well as the impact of dynamic external factors. Based on their analysis, the authors develop a framework to guide projects through the phases of planning, operationalizing, and maintaining improved decentralized energy systems in rural areas. The authors stress that there is no standard solution for such systems; the specific local context and conditions play a key role and absolutely must be considered when planning and implementing a project. In this sense, Jain and Kattuman support our plea for more comprehensive approaches that address the systemic challenges of sustainable energy solutions in developing countries.

**Acknowledgments** We would like to thank the four main authors of Part III and their co-authors for their kind, constructive, and highly motivated collaboration. On behalf of the entire authors' collective of Part III, we would also like to thank Marlène Thibault of the Centre for Development and Environment (CDE) for her patient, accurate, and highly professional editing of the texts, as well as Simone Kummer, also of CDE, for layout and graphical support in Chap. 11. Last but not least, we would like to thank Eileen Hazboun of the UNESCO Chair in Technologies for Development at the EPFL for her kind support and guidance.

## References

Douthwaite, B. (2002). *Enabling innovation: A practical guide to understanding and fostering technological change*. London: Zed Books.

Korten, D. C. (1980). Community organization and rural development: A learning process approach. *Public Administration Review, 40*(5), 480–511.

Modi, V., McDade, S., Lallement, D., & Saghir, J. (2006). *Energy and the millennium development goals*. New York: Energy Sector Management Assistance Programme, United Nations Development Programme, UN Millennium Project, and World Bank.

Mogaka, V., Ehrensperger, A., Iiyama, M., Heim, E., & Gmuender, S. (2014). Understanding the underlying mechanisms of recent *Jatropha curcas* L. adoption by smallholders in Kenya: A rural livelihood assessment in Bondo, Kibwezi and Kwale districts. *Energy for Sustainable Development Journal*, 18, 9–15.

Nussbaumer, P., Bazilian, M., & Modi, V. (2012). Measuring energy poverty: Focusing on what matters. *Renewable and Sustainable Energy Reviews, 16*(1), 231–243.

Open Working Group on Sustainable Development Goals. (2014). Sustainable Development Goals. New York.

Practical Action. (2013). *Poor people's energy outlook: energy for earning a living. [S.l.]: Practical Action Publishing*. http://cdn1.practicalaction.org/5/1/513f47d0-1950-4f85-a40f-191d0ae4f5bb.pdf. Accessed February 18, 2015.

Romijn, H., Raven, R., & de Visser, I. (2010). Biomass energy experiments in rural India: Insights from learning-based development approaches and lessons for Strategic Niche Management. *Environmental Science and Policy Journal, 13*(4), 326–338.

Schot, J., & Geels, F. W. (2008). Strategic niche management and sustainable innovation journeys: Theory, findings, research agenda, and policy. *Technology Analysis and Strategic Management Journal, 20*(5), 537–554.

Secretary-General's High-Level Group on Sustainable Energy for All. (2012). *Sustainable energy for all: A global action agenda. Pathways for concerted action toward sustainable energy for all*. http://www.se4all.org/wp-content/uploads/2014/01/SEFA-Action-Agenda-Final.pdf. Accessed February 18, 2015.

Uphoff, N., Esman, M. J., & Krishna, A. (1998). *Reasons for success: Learning from instructive experiences in rural development*. West Hartford: Kumarian Press.

Wiesmann, U., Ott, C., Ifejika Speranza, C., Kiteme, B. P., Müller-Böker, U., Messerli, P., & Zinstag, J. (2011). A human actor model as a conceptual orientation in interdisciplinary research for sustainable development. In U. Wiesmann & H. Hurni (Eds., with an international group of co-editors), *Research for sustainable development: Foundations, experiences, and perspectives. Perspectives of the Swiss National Centre of Competence in Research (NCCR) North-South*, University of Bern, (Vol. 6, pp. 231–256). Bern, Switzerland: Geographica Bernensia.

# Chapter 8
# Techno-Economic Feasibility of Green Charcoal Production in Kenya

**Kevin S. Kung, Samuel Wanderi Rigu, Steve Kariithi Karau, Kamau Gachigi and Libby McDonald**

**Abstract** Many people in emerging markets use solid fuels such as charcoal for domestic cooking. This has alarming negative environmental and economic impacts. Efforts have been made to find ways of replacing wood charcoal with waste-derived briquettes also known as "green charcoal." This chapter explores the technology, economics, and implementation of charcoal briquettes made from pyrolyzed organic waste, based on a case study in Kenya. Given the lack of formal and centralized waste management systems in emerging markets, we focus on a low-cost thermal treatment system for producing the briquettes. First, we present an economic analysis of the current domestic cooking fuel consumption pattern. The low-income households we surveyed spend a significant fraction (about 25 %) of their total income on charcoal, and the most important criterion by which they assess charcoal quality is its energy density. This is worth considering when developing alternative products. We then explore the low-cost processes enabling the conversion of agricultural residues into substitute charcoal. We find that the

K.S. Kung (✉)
Department of Biological Engineering, Massachusetts Institute of Technology, Cambridge, USA
e-mail: kkung@mit.edu

S.W. Rigu
Department of Agribusiness Management, Safi Organics, University of Nairobi, Nairobi, Kenya

S.K. Karau
School of Human and Social Sciences, Daystar University, Nairobi, Kenya

K. Gachigi
Department of Mechanical and Manufacturing Engineering, University of Nairobi, Nairobi, Kenya

L. McDonald
Department of Urban Studies and Planning, Massachusetts Institute of Technology, Cambridge, USA

 
S. Hostettler et al. (eds.), *Sustainable Access to Energy in the Global South*,
DOI 10.1007/978-3-319-20209-9_8

uptake of green charcoal production practices is possible at the community level, but depends critically on how well the product meets customer needs—and, hence, on the economic viability of the business proposition. At national and international scales, uptake tends to be enhanced by favorable guiding policies.

## 8.1 Introduction

About 2.5 billion people worldwide fuel their cooking stoves with solid biomass such as firewood or charcoal. In Kenya and many other regions of the world, charcoal production has been banned by law, partly because it is perceived as a driver of deforestation. But in urban areas, charcoal is still widespread: In Nairobi, for example, up to 700 tons of charcoal are traded daily (Njenga et al. 2013). The illegality and scarcity of charcoal in Nairobi represents an extra economic burden for the urban poor.

At the same time, proper organic waste management and disposal remain a considerable challenge in emerging markets such as Kenya. Traditionally, unlike recyclables such as plastic and metals, organic waste has very little perceived value. Even though technologies such as anaerobic digestion and composting have met with some success (e.g., Ambulkar and Shekdar 2004), they are frequently limited to source-separated urban waste because they require a steady and often sizeable supply of such waste and are frequently sensitive to feedstock variations (e.g., Klimiuk et al. 2010). In India, as an illustration, only 9.3 % of the country's organic waste is being managed by one of the technologies above (Annepu 2012). Much of the organic waste—especially agricultural residues in peri-urban farmlands—is set on fire, which creates significant environmental and health hazards (Annepu 2012).

Small-scale conversion of organic waste into "green" charcoal briquettes by means of pyrolysis—i.e., thermochemical treatment—simultaneously addresses organic waste management and domestic cooking needs. For this reason, we have investigated its potential as a complementary technology in Kenya. Our aim was to assess the feasibility and implementation of low-cost, decentralized thermal treatment systems that at the same time have the potential for commercialization and dissemination. We start by describing typical profiles of potential green charcoal users based on survey results, thereby clarifying the market and consumers' needs. Then, we discuss the relevant green charcoal technologies involved. Finally, based on interviews with existing green charcoal producers, we integrate our findings at the customer level with the broader set of social and economic factors that support or hinder the uptake of green charcoal.

## 8.2 Household Cooking Fuel Economics

Our first task was to understand whether there is a demand for solid cooking fuel in urban areas, and if so, what households' preferences are. This knowledge can provide valuable insights concerning the design of alternative products and projects for producing and marketing them.

### *8.2.1 Methodology*

We developed a questionnaire for low-income households in an urban neighborhood in Nairobi and interviewed 90 households regarding their level of income, the cooking fuel used, and their cooking habits and preferences. While the limited space in this chapter does not allow us to list the questions, they are available upon request. The survey was approved by the appropriate Institutional Review Board and adhered to its confidentiality protocols. The results in Sect. 8.2.2 are aggregated from anonymized household data. The interviewees were selected via informal networking and random sampling.

### *8.2.2 Results*

Of the participants interviewed, 67 % were females and 33 % were males. This does not reflect the population's composition but rather the fact that cooking and fuel matters are often deferred to women. Anecdotally, there were many households in which we approached a man but were then referred to a woman who was considered better placed to answer our questions. Due to the paucity of the data, we did not disaggregate the results by gender in Fig. 8.1. The interviewees had a mean daily income of US$ 3.52 ± 2.38. When asked where they do their cooking, 54 households reported that they cook indoors, 14 cook outdoors, and one switches depending on the type of fuel used; 21 left the question unanswered. Further questions revealed that smoke-emitting fuels such as firewood are typically used outside, whereas charcoal or kerosene is typically used inside. This finding is relevant: It indicates that indoor air pollution is an important factor to consider when designing a cooking fuel intervention. As most households cook indoors, the introduction of a cheaper but smokier fuel might increase the risk of long-term respiratory illnesses from indoor smoke inhalation (see e.g., Raiyani et al. 1993). Next, we proceeded to quantify the households' cooking fuel consumption pattern. Figure 8.1 breaks down the cooking fuels by type. As Fig. 8.1a shows, charcoal and paraffin[1] are by far the most common type of cooking fuel used in this

[1]We use the term "paraffin" synonymously with kerosene as a fuel.

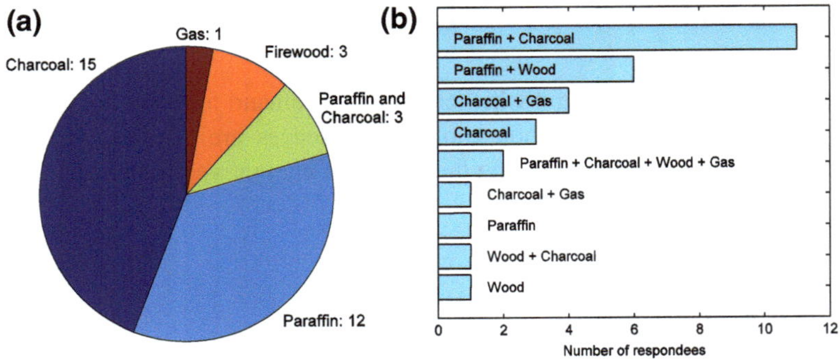


**Fig. 8.1** Fuel consumption pattern in a low-income neighborhood in Nairobi. **a** Main type of fuel used ($n = 34$ households). **b** Combination of fuels used ($n = 29$ households). The term "paraffin" is used synonymously with kerosene as a fuel; *Mawe* refers to an alternative type of briquettes made from charcoal dust and clay

neighborhood, covering about 90 % in this market, whereas firewood and gas are much less common. However, more detailed probing revealed that the fuel consumption pattern is actually much more nuanced: Households typically use more than one type of cooking fuel, switching fuel from meal to meal depending on the cooking time, ingredients, etc. Figure 8.1b quantifies this mixed fuel use in greater detail. Notably, there is significant overlap between charcoal and paraffin. Indeed, anecdotal accounts indicate that these two types of fuel complement each other: While charcoal is suitable for slow burning (e.g., stews), paraffin is preferable in situations where time is a constraint (e.g., morning tea). Much less overlap was found between firewood and gas. This is not surprising: Firewood is the cheapest fuel available, whereas gas is only accessible to those who can afford its higher price. Households using firewood probably belong to very different income categories than those who can occasionally afford gas.

A next set of questions concerned the amount of money that households spend on the different types of cooking fuel. Because these expenditures change stochastically from day to day, we estimated it by asking households to recall how much they spent on cooking fuel in the previous three days. The results are shown in Fig. 8.2. The total fuel expenditure averages about 43 % of the mean total household income, which is a surprisingly high share. Indeed, some interviewees in Nairobi told us the cooking fuel price was so high that they were occasionally forced to cut down the number of meals per day. Charcoal accounts for the greatest share of the total fuel expenditure, occupying about a quarter of an average household's total income.

Figure 8.3 shows the qualities required of charcoal by households who regularly use this type of fuel. Interviewees were asked what qualities charcoal had to have in order for them to purchase it. Surprisingly, even though households spend tremendous amounts on charcoal, the cost is of fairly low importance to them. This is likely because they have little choice in a tight market where retailers have little

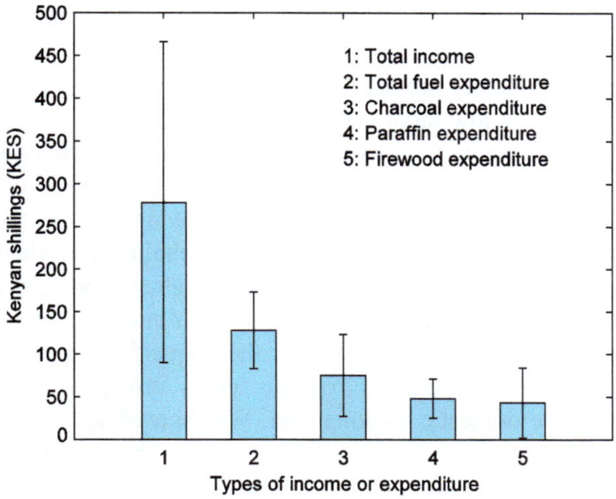


**Fig. 8.2** Daily household expenditures for different types of cooking fuel, compared to total household income

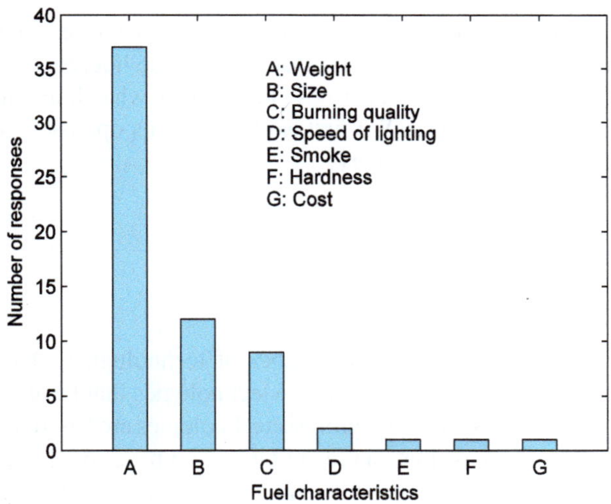


**Fig. 8.3** Qualities that interviewed households require of the charcoal they buy

room for maneuver. End users' overwhelming priority seems to be the weight of the charcoal pieces, followed by their size. This was corroborated anecdotally when we produced some alternative charcoal briquettes for the households to test. The first reaction of almost all households was to weigh the briquettes with their hands and comment on how heavy or light the briquettes felt. In retrospect, if we assume that the value of fuel is proportional to its energy content, the weight of briquettes is indeed an indirect reflection of their energy content: The denser the charcoal, the more energy it tends to contain, and this is directly reflected in the weight per volume of the charcoal pieces.

The results of this survey provided a number of key lessons. First, cooking fuel consumption patterns in low-income households in our community of interest tend to be complex, with different fuels used in different contexts, although charcoal (and paraffin) is clearly the most prominent fuel here. Second, low-income households spend a significant portion of their income on cooking fuel, to the point that it causes significant economic distress. Third, households overwhelmingly prefer heavier (and sometimes larger) charcoal pieces to lighter (and sometimes smaller) ones. The weight of the charcoal pieces reflects their energy content, and this fact should be taken into account when marketing the briquetting projects to supplement wood-derived charcoal. The findings from our survey strongly justify the search for a more cost-effective domestic cooking fuel. Widespread adoption of such a fuel and the resulting reduction in fuel expenditures could have a substantial positive impact on low-income households' energy budget.

## 8.3 Low-Cost Technologies

Based on the economics of domestic cooking fuel demand described above, this section presents promising technologies that can provide households with charcoal briquettes produced from waste at a cost lower than what households currently spend on conventional charcoal. We first explore various open-source kiln designs and then assess briquetting machinery.

### *8.3.1 Pyrolysis Kilns*

Pyrolysis can be achieved through a number of technologies. Our study mainly focuses on small-scale, low-cost, and simple technologies that have a high potential for broad dissemination. Nearly, all pyrolysis technologies are based on the following two ingredients: dry biomass that can readily be set on fire and releases enough heat for the pyrolysis to take place, and sufficient control of air supply to ensure that the biomass is turned into char as opposed to ash. We describe three technologies below that are representative of others developed by other projects in different contexts. The characteristics of these three representative designs are summarized in Table 8.1. We consider all three to be viable small-scale technologies.

The first type was developed by D-Lab of the Massachusetts Institute of Technology (MIT), in collaboration with the partners in Haiti and later elsewhere (Banzaert 2013; D-Lab 2013). This kiln consists of a 200-l oil drum with an opening at the top as well as several openings at the bottom from which the biomass is lit (bottom-lit design). Pyrolysis takes about 20 min, yielding 2 kg of charcoal dust per batch. A kiln of this type costs about US$20. The main disadvantage of this

**Table 8.1** Characteristic pyrolysis kiln designs and their properties

| | D-Lab | Takachar | Mwea Community |
|---|---|---|---|
| Capital cost (US$/unit) | 20 | 30 | 25 |
| Capacity (kg out/batch) | 2–5 | 2–5 | Up to 50 |
| Pyrolysis time (min) | 20 | 20 | 60–120 |
| Smoke emissions | High | Medium | Medium to high |
| Feedstock density | Loose | Loose | Dense |

design consists of the copious amounts of smoke that may be released during pyrolysis.

The second design is top-lit and was developed by Takachar to reduce the smoke released during pyrolysis (Kung et al. 2013). This kiln is based on a 200-l oil drum concept and modeled after other top-lit designs developed by Appropriate Rural Technology Institute (ARTI) in India and Tanzania. It has an adapter and a chimney that serve to direct the smoke upward. In addition, in this design, the biomass is lit from the top, after which the flame eventually reaches the bottom of the kiln. This kiln also takes about 20 min to perform the pyrolysis, but achieves a slightly better yield than the bottom-lit design. The advantage of this design is its reduced smoke emission—at least under windless conditions and given sufficiently dry feedstock. A kiln of this design costs about US$30. Figure 8.4 shows a sample kiln as well as different stages of its operation in the field.

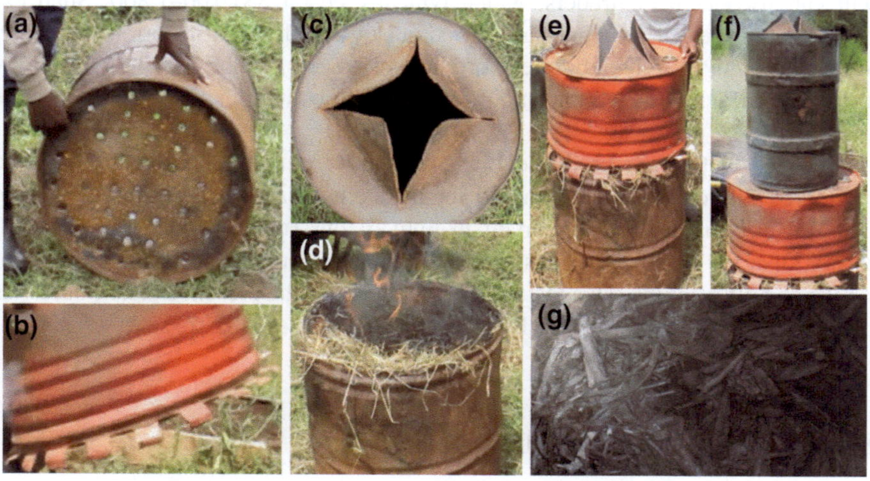


**Fig. 8.4** Carbonization process using a top-lit kiln. **a** The main kiln has an open top and many small holes at the bottom. The adapter has an alternatingly ridged/grooved bottom (**b**) as well as a star-shaped opening at the top (**c**). To operate the kiln, first set the biomass on fire from the top (**d**), and once the fire has spread all around, cover the main kiln with the adapter (**e**), immediately followed by the chimney (**f**). After about 20 min of pyrolysis followed by about 1–2 h of cooling, the charred biomass is recovered (**g**). Figure reproduced with permission from Kung et al. (2013)

The third design is used by an agricultural group in the community of Mwea and is suitable for densely packed materials such as rice husks, which may be difficult to carbonize using the two previous designs. This kiln consists of a chimney with holes in its wall, which is lit with easily flammable biomass. More densely packed material is then piled around the chimney so that all the holes are blocked. Pyrolysis by this method takes longer (up to 1–2 h), but it also yields more per batch. A kiln of this type costs about US$25.

Further pyrolysis technologies are available for other types of feedstock (such as sawdust), and some designs with larger-scale drums (e.g., by ARTI-Tanzania) are capable of uniformly carbonizing a larger batch of materials. A uniform process and design are conducive to effectively scaling the process. Technical knowledge regarding the relationships between the different technologies and different feedstock types should be pooled to facilitate effective training.

### *8.3.2 Briquetting Technologies*

There are also various types of briquetting technologies available. The lowest-cost briquette press is a simple manual device developed by MIT's D-Lab. It consists of three metal parts welded together, and its production in the informal manufacturing sector costs about US$20 per unit. When operated full time, it can produce about 40 kg of charcoal briquettes daily. More expensive hand-operated presses are also available and cost around US$ 70–250. However, larger-scale briquette production will call for electric extruders, which typically cost US$1,100–1,800. Such extruders require electricity, but have an output of 500–1000 kg of briquettes daily. When investing in a green briquette production process, the briquetting technology should be chosen to match the amount of available waste; this will ensure effective operation.

Once briquettes are produced, the emissions from these briquettes must be monitored with due diligence to avoid any inadvertent harm to the households using them. This is particularly important in light of our survey results above, which showed that the majority of households cook indoors and are susceptible to indoor smoke and other toxins. This matter is not the focus of the present chapter, but it has been addressed elsewhere (e.g., Banzaert 2013; Kung et al. 2013).

## 8.4 Commercial and Scaling Landscape

In Sects. 8.2 and 8.3, we focused on the perspective of (potential) end users of green charcoal briquettes and gave an overview of available green charcoal technologies that can be used to meet end users' needs. In this section, we assess experiences of

**Table 8.2** Socioeconomic factors promoting or hindering the uptake of green charcoal production technologies in Kenya

| | Social | Economic | Cultural | Institutional |
|---|---|---|---|---|
| Local | Lock-ins and existing relationships | Fuel economics | Prior perception versus branding | Fuel supply chain |
| Subnational | Waste management | Income generation and job creation | Tourism and ecology | Permits and licenses |
| National | Vision 2030 or similar plan for development | Income generation and job creation | Tourism and ecology | Policies and regulations |
| International | MDGs or similar development goals | International charcoal trade | "Dirty" fuel | Large-scale rollouts |

existing green charcoal enterprises in order to understand, at a higher level, the various social and economic factors that may work toward or against the uptake of green charcoal practices in Kenya.

### 8.4.1 Methodology

We conducted a survey among ten existing green charcoal enterprises operating in a variety of contexts in Kenya (for example, urban and rural). Interviews followed a questionnaire that focused on the nature of the interviewees' business operation, as well as what they perceived to be their biggest opportunities and challenges.[2] The survey was approved by the appropriate Institutional Review Board and adhered to the related confidentiality protocols. The results in Table 8.2 are aggregated from anonymized enterprise interview data. The interviewees were selected via informal networking and random sampling.

### 8.4.2 Results

Table 8.2 summarizes the experiences of several prominent Kenyan green charcoal operations regarding factors influencing the viability and scalability of green charcoal production technologies. The factors are broken down by scale of intervention (local to international) and by type of factor (social, economic, cultural, and institutional).

At the **local community scale**, the mainstay of an operation's viability is consumers' willingness to pay. This is governed by three key factors. The first is

[2]The limited space in this chapter does not allow us to list the questions. But, we are happy to provide them upon request.

fuel economics: at what energy-density-normalized price can the operation sell its charcoal briquettes, and how does this price compare to that of conventional wood-derived charcoal? The second factor is the cultural perception of briquettes: Various interviewees indicated that, in some communities, haphazard briquette-making without rigorous standards had damaged the reputation of alternative briquettes, causing people in these communities to be wary of any new products. This is a barrier to market entry that should not be underestimated. The third factor is existing relationships: Households frequently buy cooking fuel from a nearby community vendor whom they know and trust. Besides trust in the product quality, this frequently also offers them the possibility to buy fuel on credit when cash is short. Community vendors, in turn, are supplied by trusted middlemen and other well-established stakeholders in the charcoal supply chain. Gaining entry into this trust-based supply chain has proved very difficult for the most nascent green charcoal operations.

At the **subnational and national scales**, green charcoal production often encounters mixed reactions. On the one hand, it is often lauded for its potential to manage waste, improve the environment, and—most importantly—create new jobs. All these benefits are in line with the government's objectives. At the national level, for example, green charcoal production aligns with the Government of Kenya's *Vision 2030* (Government of Kenya 2011). Likewise, sustainable charcoal sourcing has occasionally even been used to promulgate ecologically responsible tourism—another goal of the national government. On the other hand, blanket regulations that are applied haphazardly and without careful consideration of the local context can and do undermine existing green charcoal production units (Bailis 2005). In recent years, with the devolution of power from the central government to the provincial governments, we expect to see subnational stakeholders take a more active role in defining the green charcoal industry.

At the **international level**, finally, things take on quite a different perspective. Green charcoal production in Kenya has two types of international audience: commercial charcoal traders and development organizations. Both are generally interested in green charcoal production. Charcoal traders from other countries, for example in the Middle East and North Africa, quite frequently contact green charcoal operators in Kenya, although a detailed assessment of this aspect is beyond the scope of our study. International development organizations often view green charcoal production as a way of contributing to international objectives such as the millennium development goals. This has led to substantial donor-driven funding and large-scale rollouts, with a mixed effect on the local charcoal industry. On the one hand, the injection of (equity-free) finance into green charcoal production enables rapid expansion and scaling of such enterprises. On the other hand, donor-funded projects sometimes overlook essential business fundamentals: They may receive funding for years without attracting real customers or making a significant social or environmental impact beyond simply providing employment for a limited number of employees involved in the project. Starting a business means a great risk for green charcoal producers, and since investments or loans are not always available, many of them opt for donor-driven funding as the safest strategy.

One other point of consideration is that charcoal in many parts of the world is perceived as a “dirty” fuel. This is only true if the trees cut down to produce the charcoal are not replanted. Alternative technologies such as anaerobic digestion or compost are often given preference over charcoal; but these solutions have their own fair share of context-dependent market and adoption challenges and are generally by no means easier to implement than green charcoal.

## 8.5 Discussion

In this chapter, we outlined the economic impact of cooking fuel consumption on low-income households and assessed the viability of green charcoal production from biomass waste. We then discussed technologies for converting unmanaged organic waste into carbonized briquettes, which have the potential to replace traditional wood-derived charcoal as a domestic cooking fuel of choice. Finally, we looked at different green charcoal production enterprises, generalizing the socio-economic factors that influence the viability and scalability of green charcoal production at different levels.

### *8.5.1 Summary of Key Lessons*

Several key lessons can be drawn from commercialization strategies pursued to date. Households prefer high-quality (i.e., heavy), high-priced briquettes, which they perceive as long-lasting, enabling them to cook more meals. However, the price per perceived weight unit is the most important factor in product acceptance. By contrast, hotels and industries place less emphasis on quality and more on a low price; therefore, any briquettes targeted toward this market segment must be able to compete with the current wholesale price of wood charcoal. Moreover, the producer must be able to deliver to the hotels and industries, as this is the current practice. Ultimately, prices are often higher in urban markets than in rural ones due to both the higher demand and the scarcity of trees nearby; this makes urban markets easier to enter. For many enterprises, industrial and institutional customers make up a large portion of the market. This seems to suggest that large institutions such as hotels and lodges may be easier to approach as potential charcoal briquette customers compared to households. Indeed, the costs of marketing a new brand of briquettes among untested households can be significant before any widespread adoption is achieved.

Green charcoal technologies are only useful if they address end users’ pain points. As shown in Sect. 8.2, the low-income households we surveyed currently spend a significant portion of their total income on cooking fuel. The central question is: Can green charcoal technologies do better than that? The pricing of green charcoal briquettes varies drastically from location to location, and based on our survey among green charcoal enterprises, transportation cost is likely the biggest factor. Nonetheless, the existence of these green charcoal enterprises shows that the solutions they offer can compete with conventional offers in terms of price, at least in their given contexts. In three cases, the price of green charcoal briquettes is at 80 % of the price of wood-derived charcoal, thus enabling the households to save 20 %.

The discussion in the previous paragraph explains the key lessons for how a green charcoal operation can take root within a community. A different set of factors governs an operation's scalability across the country. These social and economic factors are summarized in Table 8.2 in Sect. 8.4.2. Being aware of these factors and designing scaling strategies accordingly can increase the likelihood of any green charcoal enterprise to scale their operation beyond the initial set of customers.

### 8.5.2 Commercialization Strategy

Two key facts should be considered when devising a commercialization or scaling strategy. First, kilns are cheap to invest in and operate (e.g., US$20–30 apiece); this makes them suitable for distribution to various groups and individuals living or working in close proximity to organic waste (e.g., farmers, food stands, and markets). This approach of bringing the kilns to the waste is preferable to bringing the organic waste to a central site for carbonization, as charcoal dust is much cheaper to transport than the bulky waste. Second, high-quality briquettes can only be produced with expensive briquette presses. Many groups who can afford to own and operate a pyrolysis kiln cannot afford such an expensive press. For this reason, we propose a micro-enterprise model where micro-entrepreneurs collect charcoal dust from various groups or individuals and produce the briquettes at a central location. Centralized briquetting also ensures quality control as well as uniform branding and marketing—which, in turn, increases the likelihood of product acceptance (Fig. 8.5).

As shown in Fig. 8.5, the proposed setup consists of the following key features:

1. Decentralized pyrolysis: Kilns are distributed or leased to individuals or groups living in close proximity to organic waste, so that they can carbonize the waste into char at almost no cost other than labor. By selling the char to a briquetting facility, these groups or individuals can generate income from organic waste that would otherwise remain unmanaged.
2. Collection of char: The char is then collected and brought to a central briquetting facility.
3. Briquetting facility: The briquetting facility processes at least 500 kg of charcoal dust daily into high-quality and branded briquettes for distribution in urban areas.
4. Marketing and sale: Anchors in specific communities and among specific customers will assist in the sale of such briquettes.

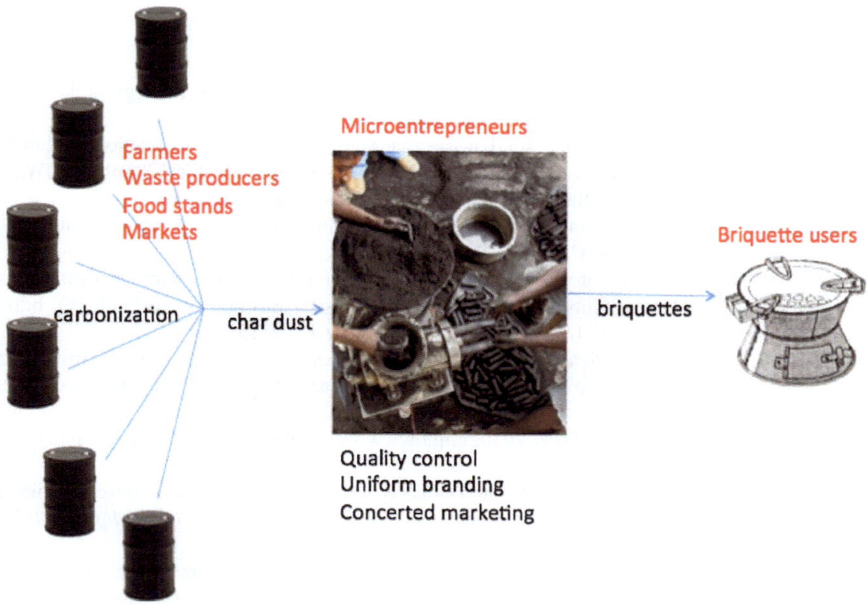


**Fig. 8.5** Proposed micro-enterprise model that combines decentralized pyrolysis with centralized briquetting and marketing

We envision each micro-enterprise as consisting of a briquetting facility and a network of kiln operators who supply the charcoal dust. A city such as Nairobi or Dar es Salaam can easily support at least five to ten such operations. Each micro-enterprise can support one micro-entrepreneur as well as three to four hired workers, in addition to providing some additional income to the network of farmers and waste-producing groups who pyrolyzed waste and supply charcoal dust. As this mode of production grows and spreads, it will be important to create a conducive environment for the proper startup and certification or commissioning of such groups at new sites by mobilizing resources and key partnerships.

**Acknowledgments** We would like to thank Jacob Young for collecting essential data and conducting interviews; Yvonne Wangare for designing and testing the top-lit drum kiln in the context of Kenya; the Zulu Youth Group and the Carolina for Kibera staff for community interviews in Nairobi, Kenya; Ali Kamil for facilitating interviews with charcoal enterprises; staff from the MIT D-Lab as well as Prof. Wanjiru Gichuhi for survey feedback and research protocol approval; and MIT Public Service Center, MIT International Development Initiative, D-Lab Scale-Ups, MIT IDEAS/Global Challenge, Legatum Center, as well as IDEO for financial support.

## References


Ambulkar, A. R., & Shekdar, A. V. (2004). Prospects of biomethanation technology in the Indian context. *Resources, Conservation and Recycling, 40*(2), 111–128.

Annepu, R. K. (2012). Sustainable solid waste management in India (Dissertation, Columbia University). http://www.seas.columbia.edu/earth/wtert/sofos/Sustainable%20Solid%20Waste%20Management%20in%20India_Final.pdf. Accessed February 18, 2015.

Bailis, R. (2005). The social and environmental implications of the charcoal trade in Kenya (Dissertation, University of California-Berkeley).

Banzaert, A. (2013). Viability of waste-based cooking fuels for developing countries: Combustion emissions and field feasibility (Dissertation, Massachusetts Institute of Technology). http://dspace.mit.edu/handle/1721.1/79158. Accessed February 18, 2015.

D-Lab. (2013). *Fuel from the fields charcoal. D-Lab charcoal process*. Massachusetts Institute of Technology. http://d-lab.mit.edu/sites/default/files/CharcoalProcess_DoIt.pdf. Accessed January 14, 2015.

Government of Kenya (2011). *Vision 2030*. http://www.vision2030.go.ke/. Accessed January 14, 2015.

Klimiuk, E., Pokoj, T., Budzynski, W., & Dubis, B. (2010). Theoretical and observed biogas production from plant biomass of different fibre contents. *Bioresource Technology, 101*(24), 9527–9535.

Kung, K., Kamil, A., Ratti, C., & McDonald, L. (2013). *Low-cost production of charcoal briquettes from organic waste*. Presented at the 2013 International Solid Waste Association Congress, Vienna, Austria.

Njenga, M., Yonemitsu, A., Karanja, N., & Ilyama, M. (2013). Implications of charcoal briquette produced by local communities on livelihoods and environment in Nairobi-Kenya. *International Journal of Renewable Energy Development, 2*(1), 19–29.

Raiyani, C. V., Shah, S. H., Desai, N. M., Venkaiah, K., Patel, J. S., Parikh, D. J., & Kashyap, S. K. (1993). Characterization and problems of indoor pollution due to cooking stove smoke. *Atmospheric Environment, 27*(11), 1645–1655.

# Chapter 9
# Putting the End-User First: Towards Addressing Contesting Values in Renewable Energy Systems Deployment for Low-Income Households—A Case from Likoma Island, Malawi

**Collen Zalengera, Richard E. Blanchard and Philip C. Eames**

**Abstract** This paper focuses on barriers and opportunities for the adoption of modern renewable energy technologies by low-income households in the light of energy requirements for household well-being; household prioritization of energy services; household purchasing power for energy; and households' experiences with traditional and conventional energy sources. Approaches for addressing the identified barriers to the adoption of renewable energy solutions by low-income households are discussed as well. Our findings show that low-income households will continue to use traditional and conventional energy sources and technologies unless modern energy solutions come with integrated financing mechanisms that enable households to pay for good-quality systems capable of meeting their energy needs. The work is based on empirical evidence from Likoma Island in Malawi.

## 9.1 Background

The positive correlation between modern energy usage and living standards (Eggoh et al. 2011; Jumbe 2004) is compelling policymakers worldwide to promote the replacement of traditional energy sources such as kerosene and firewood, which are

C. Zalengera (✉)
Mzuzu University, Luwinga, Malawi
e-mail: czalengera@yahoo.com; zalengera.c@mzuni.ac.mw

R.E. Blanchard · P.C. Eames
Centre for Renewable Energy Systems Technology, Loughborough University, Loughborough, UK
e-mail: R.E.Blanchard@lboro.ac.uk

P.C. Eames
e-mail: P.C.Eames@lboro.ac.uk

 
S. Hostettler et al. (eds.), *Sustainable Access to Energy in the Global South*,
DOI 10.1007/978-3-319-20209-9_9

**Table 9.1** Feed-in tariffs introduced by the Government of Malawi for renewable energy systems of 500 kW and more

| Renewable energy source | Feed-in tariff for | |
|---|---|---|
| | Firm power[a] (US$/kWh) | Non-firm power[b] (US$/kWh) |
| Photovoltaic | 0.20 | 0.10 |
| Biogas/biomass | 0.10 | 0.08 |
| Wind | 0.13 | |

[a]Firm power means that the producer can guarantee feeding non-intermittent power into the grid
[b]Non-firm power means that the producer cannot guarantee feeding non-intermittent power into the grid

still widely used in developing countries. This trend is further encouraged by global efforts to mitigate climate change. As a consequence, country governments are adopting a range of policies and planning measures to promote renewable energy technologies. But these efforts do not always bear fruit; in some cases, adoption of renewable energy systems does not take place at all, and in other cases, beneficiaries abandon these systems after a short time and revert to using traditional fuels. Successful adoption is underpinned by relative advantages of the available choices measured in economic terms, social prestige, convenience, and satisfaction, as well as the compatibility of available choices with existing values, past experiences, and potential adopters' needs (Rogers 1995, 2001).

In 2012, Malawi introduced feed-in tariffs as an incentive for independent power producers to invest in renewable power generation. Among other things, the government hoped that the feed-in tariffs would eventually lead to the replacement of existing diesel-fired electricity production on Likoma Island with renewable energy solutions (Malawi Energy Regulatory Authority 2012). The feed-in tariffs were set as shown in Table 9.1.[1]

Zalengera (2015) modelled a photovoltaic and wind-based energy system which could replace the existing diesel generators on Likoma Island. This model shows that the cost of producing photovoltaic and wind-based electricity could be between US$0.296 and US$0.626 per kWh, depending on the interest on capital finance, which is considerably less compared to US$0.888 per kWh for diesel-fired electricity. In 2014, the (subsidized) grid electricity tariff in Malawi averaged US$0.085 per kWh for all households, including low-income households. Thus, photovoltaic and wind systems could reduce the subsidy spent by the government on grid electricity on Likoma Island from US$0.803 per kWh to at least US$0.541 per kWh. The same study by Zalengera (2015) also showed that household-based photovoltaic and wind energy systems are more economically competitive than the existing diesel-fired electricity on Likoma Island.

The present paper examines factors affecting the deployment and diffusion of modern renewable energy technologies in low-income households, taking Likoma

[1]Geothermal and small hydropower plants are also eligible for feed-in tariffs, but are not included here because of the scope of this paper.

Island in Malawi as a paradigmatic example. In addressing this overall goal, the work focuses on:

- Assessing energy needs and requirements of households;
- Determining households' energy purchasing power;
- Determining priority energy services and household satisfaction with energy services from existing technologies; and
- Discussing potential barriers and opportunities for renewable energy adoption on Likoma Island.

## 9.2 Methodology

### *9.2.1 Study Site*

The study was carried out on Likoma Island, which is located in Lake Malawi and has a land area of 18 km$^2$ (National Statistics Office 2008). The island is home to about 10,500 people living in 1500 households, which amounts to an average household size of seven people. The major sources of energy on the island are firewood for cooking, as well as kerosene and dry-cell-battery-powered torches for lighting. Electricity is supplied by diesel generators feeding a grid system that is independent of the mainland grid. Due to high operation costs, the generators are switched on only from 6 a.m. to 12 noon and from 2 to 10 p.m., that is, during 14 h each day.

### *9.2.2 Data Collection and Analysis*

We did a survey in January 2013 using a semi-structured questionnaire (Laws et al. 2002; Mikkelsen 2005) administered to households in face-to-face interviews and for self-completion. The participating households were selected using proportionate stratified random sampling (Curwin and Slater 2004; Laws et al. 2002; Mikkelsen 2005; Scheyvens and Storey 2003) based on the population of each of the twelve villages on Likoma Island. Survey participants were selected from village registers in the presence of the respective village chiefs, who were then asked to comment on the representation of different social classes in the final sample. When necessary, participants were added until the desired social class was represented. The final retained sample size was 202 households.

The household energy requirement for good well-being (Dodge et al. 2012) was estimated based on households' energy needs, their current energy consumption, and the authors' assessment of what electrical appliances are required for comfortable living. Households were asked to name their energy needs by selecting appropriate entries from a list. The needs listed were lighting, cooking, water heating, radio, television (TV), telecommunication/phone charging, refrigeration, and space cooling/water heating. Respondents were also asked about the number of rooms in their property (to determine lighting energy demand[2]), as well as their daily hot water requirement, the daily amount of firewood used for cooking, electrical appliances owned, and electrical appliances they planned to purchase. In order to determine energy purchasing power, the questionnaire elicited quantitative data on households' total monthly expenditure on energy services (excluding for transport and mobility); how much they were willing to pay per month for all household energy services (willingness to pay for energy per month); and the maximum they could afford in terms of capital costs of an energy system (willingness to pay for one-time purchase of energy system).

In addition, households were asked to rank the importance of selected household energy services such as cooking, lighting, water heating, radio, TV, and phone charging on a scale of 1–10 (1 = most important). Lastly, respondents were asked to rank their satisfaction with the aforementioned energy services on a scale of 1–4 (1 = very dissatisfied, 2 = dissatisfied, 3 = satisfied, and 4 = very satisfied).

Data analyses consistent with the data types—nominal, ordinal, or scale (Curwin and Slater 2004; Leech et al. 2008; Morgan et al. 2004)—were carried out in SPSS and Microsoft Excel. The energy expenditures on individual energy services were summed up to determine the total energy expenditure for each respondent household. This was averaged within a confidence interval of 95 % for all respondent households. Likewise, households' monthly willingness to pay for energy services and the willingness to pay for the one-time purchase of an energy system were averaged within a confidence interval of 95 %.

Importance ranks of energy services were first reversed so that what had been ranked 1 (most important) was now assigned the highest rank 10; the importance rank of each service was then averaged within a confidence interval of 95 %. The ranks for satisfaction with energy services were also averaged within a confidence interval of 95 %. The ranks of importance of each energy service and the satisfaction scores of each energy service were then summed up for all respondents and normalized to their maximum expected sums determined by the number of valid responses and the possible maximum score. This normalization helped to validate the results obtained by averaging the ranks for the priority of services and satisfaction with the services.

[2]We assumed that one energy-efficient 15-W compact fluorescent light bulb is sufficient for a room; the Malawi government distributed energy-efficient bulbs to replace incandescent bulbs in all households with access to electricity. Two additional compact fluorescent lights are required for outdoor lighting from 6 p.m. to 6 a.m.; and the lighting for toilets and bathrooms was assumed to be equivalent to one 15-W bulb operating for one hour.

## 9.3 Findings

### 9.3.1 Household Energy Requirement

**Artificial Lighting** Households require artificial lighting from 6 p.m. to 10 p.m. daily. An average living property on Likoma Island has seven rooms (three bedrooms, living room, storeroom, corridor, and veranda; a kitchen is not usually part of the main property but is included in the energy requirement calculations). Two additional compact fluorescent lights are required for outdoor lighting from 6 p.m. to 6 a.m.; and the lighting for toilets and bathrooms was assumed to be equivalent to one 15-W bulb operating for one hour.

**Electrical Appliances** Households require the following electrical appliances (power ratings estimated by the authors): a refrigerator (225 W), a colour TV screen (70 W), a satellite decoder (30 W), a radio (30 W), and five cooling fans (45 W each). A hotplate (1,500 W) is required for cooking three times a day, for a total of six hours per day (5 to 6 a.m., 10 a.m. to 1 p.m., and 6 to 8 p.m.). Water can be heated using solar thermal energy.

**TV Screens** TVS are usually used with a decoder or a VCR or DVD player for eight hours per day (6 to 7 a.m., 12 to 3 p.m., and 6 to 10 p.m.). Indeed, during fieldwork, we observed in some households that TV screens were switched on soon after waking up in the morning, remaining on through breakfast time until the household members left to go about their daily livelihood activities. During lunchtime, the TV was again switched on, and after lunch, children would continue watching films on VCR/DVD depending on whether or not they had other activities to do; based on our observations, we set 3 p.m. as the cut-off time for children's daytime TV usage to take account of the time they spend doing household chores and participating in community social activities. After 6 p.m., when all household members were at home, the TV screen usually remained on until bedtime around 10 p.m.; during these hours, there was always at least one person in the living room (usually males). In addition to TV, a radio is used for 8 to 12 h daily when the TV is switched off, although sometimes we observed both appliances to be simultaneously on.

**Cooling Fans** Households need cooling fans because temperatures on Likoma are above comfort levels as defined for Malawi by Zingano (2001). Three cooling fans (one for each bedroom) need to be on overnight (one from 10 p.m. to 6 a.m., and two from 8 p.m. to 6 a.m. for those who go to bed earlier). Fans in the other rooms are required whenever someone is there, which, based on our observations, we assumed to be 6 to 7 a.m., 12 noon to 3 p.m., and 6 to 10 p.m. for the living room, totalling eight hours daily; as well as 11 a.m. to 12 noon and 6 to 8 p.m. for the kitchen, totalling three hours daily. Toilets and storerooms do not require cooling.

**Daily Household Energy Requirement** Fig. 9.1 shows the daily household energy requirement by energy needs. Cooking contributes the largest share, followed by refrigeration, space cooling, water heating, entertainment, and lighting.

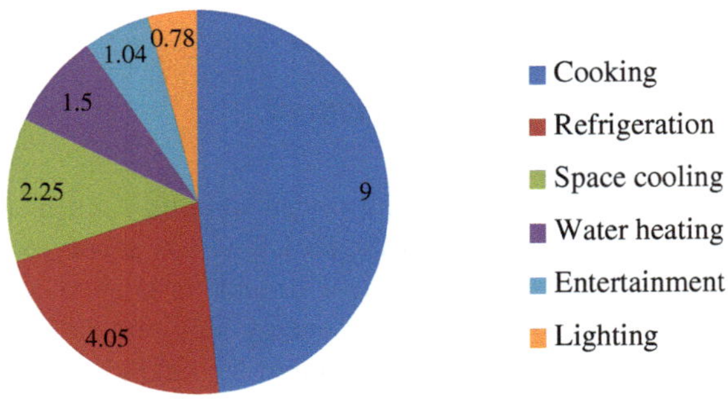


**Fig. 9.1** Distribution of household energy requirements (in kWh) by energy needs. Entertainment includes radio, TV, and VCR/DVD player. The total energy requirements amount to about 19 kWh (sum of all numbers provided in the figure)

As can be seen from Fig. 9.1, the daily household energy requirement on Likoma Island is estimated at 19 kWh. This adds up to 6796 kWh annually, which is consistent with the minimum per capita energy consumption defined by Sanchez (2010), as cited by Practical Action (2010). According to these authors, in order to live above the energy poverty line: "*A person should have access to at least: (a) the equivalent of 35 kg LPG for cooking per capita per year from liquid and/or gas fuels or from improved supply of solid fuel sources and improved (efficient and clean) cook stoves*; *and (b) 120 kWh electricity per capita per year for lighting, access to most basic services (drinking water, communication, improved health services, improved education services and others) plus some added value to local production*". For a family of seven persons, this means a minimum cooking fuel requirement of 8.5 kWh daily, assuming 46 MJ/kg for liquefied petroleum gas (LPG) and an electricity requirement for lighting of 2.3 kWh daily.

### *9.3.2 Purchasing Power for Energy*

Table 9.2 presents households' energy purchasing power. It appears that the total monthly energy expenditure is twice as high as the willingness to pay for energy during the same period. In addition, nearly 50 % of the respondent households indicated having experienced financial pressure due to energy bills.

#### 9.3.2.1 Energy Service Prioritization

Table 9.3 presents households' prioritization of energy services. Lighting is perceived as the most important energy service, followed by cooking, TV, radio, telephone, and water heating.

Table 9.2 Energy purchasing power of households on Likoma Island (US$ 1 = 400 Malawi Kwacha)

| | Total monthly expenditure on energy (US$) | Willingness to pay for energy per month (US$) | Willingness to pay for one-time purchase of energy system (US$) |
|---|---|---|---|
| Lower bound of mean | 9.00 | 5.00 | 66.00 |
| Upper bound of mean | 12.00 | 6.00 | 93.00 |
| Central mean | 11.00 | 5.28 | 80.00 |
| Median | 8.50 | 4.00 | 56.00 |
| Standard deviation of mean | 8.70 | 5.00 | 78.00 |
| Standard error | 0.68 | 0.38 | 7.00 |
| Skewedness of data | 1.345 | 2.936 | 1.484 |
| Kurtosis of data | 1.796 | 11.895 | 2.236 |
| Valid responses (from 202) | 163 | 179 | 130 |

95 % confidence interval for mean

Table 9.3 Prioritization of energy services by households on Likoma Island. Importance values range from 1 (=most unimportant) to 10 (=most important)

| How important is… | Mean | Median | Skewedness of data | Normalized sum of ranks | Valid responses |
|---|---|---|---|---|---|
| … cooking | 8.50 | 9.00 | −2.308 | 1029/1121 = 0.85 | 121 |
| … lighting | 8.93 | 10.00 | −2.608 | 1098/1230 = 0.89 | 123 |
| … water heating | 7.15 | 8.00 | −1.172 | 472/660 = 0.72 | 66 |
| … radio | 7.29 | 7.00 | −0.869 | 634/870 = 0.73 | 87 |
| … TV | 7.56 | 8.00 | −0.788 | 597/790 = 0.75 | 79 |
| … telephone | 7.17 | 7.00 | −0.440 | 502/700 = 0.72 | 70 |

Mean and median values in the table are based on these values

#### 9.3.2.2 Satisfaction with Energy Services

Table 9.4 presents households' satisfaction with current energy services. The mean satisfaction scores for all energy services are below 3, which clearly indicates that people are dissatisfied with the existing energy services.

## 9.4 Discussion

The following sections discuss barriers and opportunities for the adoption of renewable energy technologies on Likoma Island in the light of energy requirements, purchasing power, the importance attributed to different energy services, satisfaction with energy services, and the theory of diffusion of innovations.

**Table 9.4** Satisfaction scores for household energy services on Likoma Island (4 = very satisfied; 3 = satisfied; 2 = dissatisfied; 1 = very dissatisfied)

| Satisfaction with energy services for ... | Mean | Median | Skewedness of data | Normalized sum of ranks | Valid responses |
|---|---|---|---|---|---|
| ... cooking | 2.38 | 2 | 0.165 | 331/556 = 0.595 | 139 |
| ... lighting | 2.34 | 2 | 0.157 | 320/448 = 0.714 | 137 |
| ... water heating | 2.33 | 2 | 0.159 | 207/356 = 0.581 | 89 |
| ... powering radio | 2.27 | 2 | 0.28 | 207/364 = 0.568 | 91 |
| ... powering TV | 2.33 | | 0.332 | 135/232 = 0.581 | 58 |
| ... charging mobile phones | 2.64 | 3 | −0.071 | 156/236 = 0.661 | 59 |

## 9.4.1 *Barriers to Renewable Energy Development on Likoma Island*

### 9.4.1.1 Local Socioeconomic Barriers

At Malawi's current electricity tariff of US$0.085 per kWh, a household on Likoma Island that relies entirely on grid electricity for its energy requirement would spend about MWK 19,000.00 (US$47.48) on energy per month (not including water heating which can be achieved using solar thermal energy). This is four times the prevailing monthly average energy expenditure, which we estimated at US$11.00 based on our survey, and it is nine times the prevailing willingness to pay for energy, which we estimated at US$5.28 per month (Table 9.2). Thus, based on the prevailing willingness to pay for energy, households can afford only 11 % of their daily energy requirement from grid electricity. Thus, promoting grid electricity for meeting all household energy needs would put financial pressure on households or increase energy poverty. Since there is no evidence that electricity from renewable energy sources would reduce the electricity tariff (see Sect. 9.1), it is highly likely that a significant proportion of households would continue to use traditional energy technologies in order to keep their energy expenditure within their willingness to pay. Moreover, upfront investments for connection (US$70) and electrical wiring (about US$500) (Zalengera et al. 2014) exceed households' willingness to pay for the one-time purchase of an energy system.

### 9.4.1.2 National Institutional Barriers

The minimum size of systems eligible for feed-in tariffs, which was set at 500 kW, is the result of a technology-centred approach aimed at supporting affluent entrepreneurs without addressing the challenges faced by low-income households to access clean energy. Although electricity from renewable energy sources is cheaper to produce than the diesel-fired electricity, Malawi's feed-in tariffs are still relatively lower than the cost of generating renewable energy (see Sect. 9.1). This is unconducive for private-sector investment and thus contradicts the original intention behind the introduction of feed-in tariffs.

#### 9.4.1.3 Market System Barriers

In our case study on Likoma Island, we noted that developers tend to respond to households' low purchasing power by marketing cheap technologies, for example, solar lanterns and 50–100-W solar home systems. Although it is arguably useful, these market-based solar lanterns and solar home systems are not only inadequate but also focus on energy needs that do not significantly contribute to household energy demand. In addition, due to the island's remoteness, compounded by households' low income, professional after-sales support for the installed systems is usually not available—which in turn has an adverse effect on their reliability.

### *9.4.2 Opportunities for Renewable Energy Development on Likoma Island*

Considering our findings, the theory of diffusion of innovations (Rogers 1995, 2001), and the renewable energy system modelled by Zalengera (2015; see Sect. 9.1), there are a number of opportunities for renewable energy development on Likoma Island. They are outlined in the following sections.

#### 9.4.2.1 Potential for User Satisfaction and Convenience

The results in Tables 9.3 and 9.4 show that the prevailing energy technologies do not satisfy households' needs. We noted during the survey on Likoma Island that poor satisfaction with energy services is linked to people's bad experiences with traditional and conventional technologies. Households without access to electricity expressed dissatisfaction due to indoor air pollution from solid and liquid fuels, which causes coughs; the need for continuous attention to ensure that firewood keeps burning; the time spent collecting firewood and preparing the fire; backaches due to drudgery associated with firewood; the non-durability of dry cell batteries; spillage of acidic gel from dry cell batteries; as well as bad odours and the risk of accidents from kerosene lamps. By contrast, households with access to grid electricity expressed poor satisfaction with energy services due to the intermittent operation of the diesel generators feeding the grid. Households' dissatisfaction with traditional and conventional energy solutions is an opportunity for introducing renewable energy technologies that address the dislikeable properties of traditional and conventional technologies and thereby provide a real alternative for local households. A better understanding of existing dissatisfaction could be useful in planning and designing adequate alternative systems.

#### 9.4.2.2 National Economic Advantages

Renewable energy technologies have the potential to reduce the costs of electricity generation on Likoma Island—and hence government expenditure on tariff subsidy —by at least US$0.262 per kWh of electricity supplied to households. This is equivalent to a reduction of at least 33 % compared to the current subsidy. Furthermore, renewable energy technologies are compatible with international values and policies on reducing carbon emissions and could therefore attract financing from international multilateral organizations such as the United Nations Framework Convention on Climate Change (UNFCCC n.d.), through its Clean Development Mechanism (CDM), and the Global Environmental Facility (GEF 2013).

## 9.5 Conclusion and Way Forward

Given the findings and discussion presented in this paper, it is essential that energy practitioners developing technical solutions make use of the available knowledge about people's satisfaction with services, their priority energy services, and any dislikable or likeable properties of existing technologies. According to the theory of diffusion of innovations, households—including those with a low income—are more likely to pay for, and maintain, costly but reliable products if these technologies are aligned with social values and preferences, as adoption of such technologies can increase households' social prestige and is compatible with their values.

We propose a review of the Malawi feed-in tariffs that were introduced in 2012 and still present a significant barrier to low-income households on Likoma Island in accessing adequately sized sustainable renewable energy solutions. Given that household-based photovoltaic and wind systems are also potentially more economic than diesel-fired electricity, it might make sense to reduce the minimum size of renewable energy systems eligible for the feed-in tariff to the order of 1–5 kW and to introduce loans for covering capital costs which could be repaid within a period determined by the households' willingness to pay for capital costs as presented in Table 9.2. It should be noted that when households generate their own power, this is similar to feeding power into the grid; excess energy would indeed be fed into the grid. Accordingly, households could earn money from the feed-in tariff, which could contribute to repayment of capital costs and probably also address households' income poverty (Zalengera et al. 2014). In addition, since cooking is a major contributor of energy demand (see Fig. 9.1), the feed-in tariff (or equivalent incentives) and capital financing could be extended to other end-user renewable energy technologies that can be used directly for cooking, such as biogas or solar thermal cookers.

In view of market-based technologies, we agree with Lindner (2011) that a sustainable approach to disseminating renewable energy systems should not be based on attempts to increase affordability by developing cheaper products, but on fostering income-generating activities (and financing mechanisms) that enable people to pay for good-quality systems that are capable of meeting their energy needs.

**Acknowledgments** The research project to which this work is part was sponsored by the Commonwealth Scholarship Commission with funds from the United Kingdom Department for International Development.

## References

Practical Action. (2010). *Poor people's energy outlook 2010*. Rugby, UK: Practical Action.

Curwin, J., & Slater, R. (2004). *Quantitative methods for business decisions* (5th ed.). London: Thomson.

Dodge, R., Daly, A. P., Huyton, J., & Sanders, L. D. (2012). The challenge of defining well-being. *International Journal of Well-being, 2*(3), 222–235.

Eggoh, J. C., Bangake, C., & Rault, C. (2011). Energy consumption and economic growth revisited in African countries. *Journal of Energy Policy, 39*(11), 7408–7421.

Global Environment Facility [GEF]. (2013). What is GEF? http://www.thegef.org/gef/whatisgef. Accessed July 15, 2014.

Jumbe, C. B. (2004). Cointegration and causality between electricity and GDP: Empirical evidence from Malawi. *Journal of Energy Economics, 26*(1), 61–68.

Laws, S., Harper, C., Jones, N., & Marcus, R. (2002). *Research for development: A practical guide*. London: SAGE Publications Ltd.

Leech, N. L., Barrett, K. C., & Morgan, G. A. (2008). *SPSS for intermediate statistics-use and interpretation* (3rd ed.). New York, London: Psychology Press.

Lindner, K. (2011). Quality issues in the market based dissemination of solar home systems. In M. Schafer, N. Kebir, & D. Philipp (Eds.), *Micro Perspectives for Decentralized Energy Supply,* (pp. 220–229). Proceedings of the International Conference, Technische Universität Berlin, 7–8 April.

Malawi Energy Regulatory Authority. (2012). *Malawi feed-in tariff: Renewable energy resource generated electricity in Malawi*. Lilongwe: Malawi Energy Regulatory Authority.

Mikkelsen, B. (2005). *Methods for development work and research: A new guide for practitioners*. New Delhi, London: SAGE Publications Ltd.

Morgan, G. A., Leech, N. L., Gloeckner, G. W., & Barrett, K. C. (2004). *SPSS for introductory statistics—use and interpretation* (2nd ed.). Mawwah, New Jersey, London: Lawrence Erlbaum Associates Publishers.

National Statistics Office. (2008). *2008 population and housing census*. Zomba, Malawi: National Statistics Office.

Rogers, E. M. (1995). *Diffusion of innovations* (4th ed.). New York: Free Press.

Rogers, E. M. (2001). Innovation, theory of. In Neil J. Smelser & Paul B. Baltes (Eds.), *International encyclopaedia of the social & behavioral sciences* (pp. 7540–7543). doi: 10.1016/B0-08-043076-7/01900-8.

Sanchez, T. (2010). *The hidden energy crisis: How policies are failing the World's poor*. Rugby, UK: Practical Action Publishing.

Scheyvens, R., & Storey, D. (2003). *Development fieldwork: A practical guide*. London: SAGE Publications Ltd.
United Nations Framework Convention on Climate Change [UNFCC]. (n.d). What is the CDM. http://cdm.unfccc.int/about/index.html. Accessed May 24, 2014.
Zalengera, C. (2015). A study into the techno-economic feasibility of photovoltaic and wind generated electricity for enhancement of sustainable livelihoods on Likoma Island in Malawi. *PhD thesis*, Loughborough University, Loughborough.
Zalengera, C., Blanchard, R. E., Eames, P. C., Juma, A. M., Chitawo, M. L., & Gondwe, K. T. (2014). Overview of the Malawi energy situation and a PESTLE analysis for sustainable development of renewable energy. *Journal of Renewable and Sustainable Energy Reviews, 38*, 335–347.
Zingano, B. W. (2001). A discussion on thermal comfort with reference to bath water temperature to deduce a midpoint of the thermal comfort temperature zone. *Journal of Renewable Energy, 23*(1), 41–47.

# Chapter 10
# Energy Poverty and the Perception of, and Satisfaction with, Renewable Energy Technologies: The Case of Solar Villages in Pakistan

**Bilal Mirza**

**Abstract** This paper assesses users' satisfaction with renewable energy technologies (RET) and compares it with non-users' perception of RET. Non-users' perception of RET was investigated in nearly 600 households in 25 rural communities in Punjab province, Pakistan. User's satisfaction with RET was assessed in two solar villages in northern Punjab that had been equipped with solar photovoltaic (PV) systems in 2004–2005 as an alternative to on-grid electricity. Our results reveal major discrepancies between perceptions of RET in non-RET communities and users' satisfaction with solar PV systems in RET communities. Rural households in non-RET communities have extremely high expectations of RET, especially of solar PV systems; by contrast, households who had adopted solar PV systems showed high levels of resentment, anger, and disappointment toward this technology. Our findings also show that deployment of solar PV systems has not led to the expected improvements in user households' energy poverty status. Thus, merely providing appropriate technology to the poor is not enough to ensure success. A sense of ownership and acceptance of the technology among local residents is needed in order for the technology to contribute to development and, eventually, poverty alleviation.

## 10.1 Introduction

Providing energy access for rural communities remains an immense challenge in many developing countries. Many national governments, nongovernmental organizations, international donors, and also some rural households perceive renewable

B. Mirza (✉)
Centre for Policy Studies, COMSATS Institute of Information Technology, Islamabad, Pakistan
e-mail: bilal.mirza@comsats.edu.pk; mirza@merit.unu.edu

B. Mirza
United Nations University-MERIT, Maastricht University, Maastricht, Netherlands

 
S. Hostettler et al. (eds.), *Sustainable Access to Energy in the Global South*,
DOI 10.1007/978-3-319-20209-9_10

energy technologies (RET) as the panacea for all their energy problems. Numerous RET projects were implemented across sub-Saharan Africa and the South Asian subcontinent with the aims of ensuring sustainable energy supplies (Bajracharya et al. 1991; Bikam and Mulaudzi 2006), reducing energy poverty in rural areas (Huraib et al. 1996; Farhar and Houston 1996; Zein and Sarar 1998; Fara et al. 1998; Howells et al. 2005), and even overcoming the energy crisis (Farooq and Kumar 2013; Kessides 2013). However, a lack of favorable policy frameworks as well as of community involvement and local-level capacity building has had strong negative impacts on the deployment of RET in rural communities (Bikam and Mulaudzi 2006).

RET have been deployed, tested, and assessed with regard to their social, economic, and environmental benefits in many countries, especially in China, India, Nepal, and some parts of sub-Saharan Africa (Bajracharya et al. 1991; Rajabapaiah et al. 1993; Pokharel 2006). The literature hosts a mix of evidence-based opinions on whether RET contribute to improving people's energy security and eventually their livelihoods, particularly in remote rural communities in developing countries. A positive impact and success stories are reported by Alam et al. (2003), Obeng et al. (2008), Palit (2013), and de Martino Jannuzzi and de Melo (2013), whereas failures are discussed by Lemaire (2011), Bhattacharya and Jana (2009), and Millinger et al. (2012). Several studies have also assessed perceptions, user satisfaction, suitability, operational capability, and network and market factors for different RET and different types of consumers (see Gullberg et al. 2005; Bikam and Mulaudzi 2006).

Researchers have suggested that RET, apart from their economic and social benefits, could also help to alleviate widespread energy poverty (Dendukuri and Mittal 1993; Rajabapaiah et al. 1993; Biswas et al. 2001; Platt and Wilson 1999; Balinet 2006; Wamukonya 2007). Fara et al. (1998) stressed economic development, infrastructure modernization, higher degrees of education and cultural activity, professional prospects for young people, and sufficient energy resources (sunlight) as key conditions for the successful implementation of solar photovoltaic (PV) technologies in Romania. According to Biwas et al. (2001), RET can only qualify as appropriate if they (1) are easy to operate, implement, and maintain; (2) provide comfort; (3) save resources (energy, time, etc.); and (4) increase labor productivity.

Numerous research efforts were conducted by international organizations[1] in developing countries to understand the social benefits of solar and other RET. Chandrasekar and Kandpal (2007) conducted an opinion-based survey in India where they asked different stakeholders—experts, government officials, researchers,

[1]The International Centre for the Application of Solar Energy (CASE)—the United Nations Industrial Development Organization's agency for the promotion of renewable energy technologies in developing countries—has actively engaged in rural electrification projects since the early 1990s. Based on its integrated developmental approach, CASE analyzed how the supply of energy can add vigor to the economic and social life of people in rural communities of developing countries (Thompson and Singh 1996).

NGOs, and internationally funded institutions—about the status of different RET. They found that state agencies play a pivotal role in determining the future potential of RET. However, technology manufacturers can also contribute to technology development, quality control, standardization, market development, and program design. Oddly, this survey does not include the opinion of actual users of RET. Mala et al. (2009) conducted a survey to assess the usefulness of solar PV systems in the Republic of Kiribati and found in their initial conclusions that implementing such systems is not suitable as a universal strategy for improving the livelihoods of people using them. Thus, it seems essential to gain a better understanding of how rural households using these technologies perceive and assess them.

The main objective of this paper is to assess users' satisfaction with RET—or, more specifically, solar PV systems—and to compare it with the expectations of households who do not yet have access to these technologies. Following this introduction, we discuss the design and methods of our study, providing information about the study area in Punjab province, Pakistan, and the sampling technique used for our survey, which we refer to as the Energy Poverty Survey. Subsequently, we present our findings concerning nonuser households' perceptions of different RET in villages with no access to RET, as well the satisfaction with solar PV systems of user households in two villages where such systems were installed by a government project. Based on the findings, we then explore the main reasons why the implementation of solar PVs did not succeed in these two communities. The paper ends with some more general conclusions and a number of recommendations.

## 10.2 Study Design and Methodology

### *10.2.1 Study Area and Timing of Survey*

We chose to conduct our survey in Punjab province of Pakistan for several reasons. The rural population of Punjab represents the largest proportion of the country's total rural population (51 million, or 57 %, out of 89 million). Over 70 % of people in Punjab live in rural areas and earn their livelihoods directly or indirectly from agriculture. Nearly 40 % of this rural population still uses different traditional energy sources, including traditional biomass (firewood and animal, plant, and agricultural waste), kerosene, liquid petroleum gas, and electricity from diesel and gas generators and rechargeable batteries. In some communities, the national RET promotion agency (Alternative Energy Development Board, AEDB) has installed various RET, most notably solar PV systems, with the help of external donors.

The fieldwork was carried out from September 2008 to January 2009, as the fall and winter seasons were most appropriate in view of the local climate and harvest periods.

### 10.2.2 *The Energy Poverty Survey: Sampling, Main Questions, and Response Rates*

**Selection of Districts**. To select our respondents, we first drew a purposive sample of eleven districts where we expected a high proportion of rural communities without natural gas, based on data from the Pakistan Bureau of Statistics.[2] Then, we used different sampling methods to create a sample of households with no access to RET (non-RET households) and one of households with access to RET (RET households).

**Sampling of Non-RET Households**. To create our sample of non-RET households, we drew a purposive targeted sample of 25 communities in ten of the eleven districts—three at most from each district—that had no access to natural gas and, if possible, no access to electricity. Once these non-RET communities were identified, we used a random sampling method to select 30 households in each community. During fieldwork, the targeted sample of 750 households from ten districts was reduced to 730 households for various reasons.

**Sampling of RET Households**. For our assessment of users' satisfaction with RET, we selected two neighboring villages which the AEDB had equipped with solar PV systems in 2004–2005 as an alternative to on-grid electricity: Khorian and Narian, situated in the north of Punjab province, in Rawalpindi district, nearly 55 km from the district capital of Rawalpindi (Fig. 10.1). We selected these two solar villages (RET communities) because they are located within a safe district; other RET communities are located in the high-risk district of Dera Ghazi Khan and could not be considered. Both Khorian and Narian lack access to grid-based electricity and natural gas. Until the beginning of 2009, when we conducted the survey, there were neither signs of electric poles nor pipelines for natural gas, nor any infrastructural developments hinting toward potential future development. Using a random sampling method, we selected 25 households from each village, amounting to a total of 50 households using RET.

**Focus of the Energy Poverty Survey**. In our survey, apart from questions related to household income, occupation, and energy mixes, we specifically asked non-RET respondents about their knowledge and perceptions of RET. Our foremost question in this respect was whether they had heard about RET as an energy solution.

**Response Rates**. Out of the two samples totaling 780 households in 27 communities across 11 districts, 640 households responded, which amounts to a response rate of 82 %. Table 10.1 shows for each district the number of rural communities surveyed and the number of households who actually responded.

Within the Rawalpindi district, where we surveyed the two RET communities only, the response rate was an astounding 100 %, which can be explained with people's resentment against the government for failing to solve persistent energy problems.

---

[2]For more details, see Area and Population of Administrative Units report, available online at http://www.pbs.gov.pk/content/area-population-aadministrative-units.

**Fig. 10.1** View of solar PVs installed in rural households

**Table 10.1** Rural communities surveyed and number of respondent households per district

| District | | Number of communities surveyed | Number of respondent households |
|---|---|---|---|
| *Non-RET communities/households* | | | |
| 1 | Chakwal | 3 | 75 |
| 2 | Jhelum | 2 | 50 |
| 3 | Gujarat | 1 | 24 |
| 4 | Sialkot | 3 | 75 |
| 5 | Lahore | 2 | 50 |
| 6 | Sheikhupura | 3 | 75 |
| 7 | Faisalabad | 3 | 74 |
| 8 | Multan | 2 | 46 |
| 9 | Muzaffargarh | 3 | 58 |
| 10 | Layyah | 3 | 63 |
| *RET communities/households* | | | |
| 11 | Rawalpindi (solar villages) | 2 | 50 |
| | Total | 27 | 640 |

Among non-RET households, the response rate was 80 %. There are various reasons why one-fifth of the households did not participate. In rare cases, this was due to the absence of an appropriate respondent—that is, an adult household member who knew about the household's energy requirements, expenditure, and energy sources. This mostly occurred in wealthier households, whose members were living either in abroad or in another place within the country and were thus not available to respond. However, a closer analysis of non-respondents and respondents revealed that the request to fill in the survey was turned down mostly by female household members. This fact can largely be attributed to social constraints on female participation: It is uncommon for women in rural Punjab to interact with strangers. It is pertinent to mention in this context that our sample of respondents shows a clear gender bias in favor of males: Only 6.4 % of all respondents (non-RET and RET) were female. Nonetheless, we noticed differences between male and female respondents, with women showing greater awareness of household energy problems than men.

## 10.3 Results

### *10.3.1 Appliances Owned, Energy Sources Used, Income, and Occupation in RET Communities*

According to the AEDB sources, 110 households in the two selected RET communities are electrified: 57 in Khorian and 53 in Narian. In our survey, local residents reported somewhat lower figures of 35 households in Khorian and 45 in Narian. The reason for this difference might be that the AEDB reports the number of installed solar PV systems rather than the number of homes having such a system; in a few homes, we found more than one solar panel installed.

To better understand households' electricity demand, we asked them what electric appliances they own and use. The results are presented in Table 10.2. In Khorian, 100 % of the households interviewed owned electric appliances; in Narian, 95.8 % of the households did. Most households owned bulbs or tube lights (Khorian: 100 %; Narian: 95.8 %) and ceiling or pedestal fans (Khorian: 84.6 %; Narian: 91.3 %).

Table 10.3 provides an overview of the two communities regarding energy sources used, as well as income and occupation. Households meet their domestic energy needs from different energy sources, including firewood, animal waste, and kerosene. None of the households from either village reported to be using plant waste or liquid petroleum gas; the former is used as fodder for cattle and other domestic animals, and the latter is not available, as there is no distribution network. Farming is still the most common activity among households in the two villages (Khorian: 73 %; Narian: 78.2 %). The average household income is higher in Khorian than in Narian, even though Khorian has an unemployment rate of 11.5 %, whereas all respondents were employed in Narian.

### *10.3.2 Energy Poverty Among RET Households*

Table 10.4 compares selected energy poverty indicators[3] of RET and non-RET rural communities. One of these is the Energy Inconvenience Excess Index (EIE) that expresses the degree of inconvenience (per capita) in positive and negative numbers. The EIE is based on seven indicators that are relevant for energy access by rural households: frequency of buying or collecting a source of energy; distance from household travelled; means of transport used; household members' involvement in energy acquisition; time spent on energy collection per week; household health effects; and children's involvement in energy collection. A positive EIE value indicates a relative state of convenience, whereas a negative value indicates a relative state of inconvenience incurred by households while accessing their energy

[3]See Mirza and Szirmai (2010) for more information on the computation of EII, ESI, and EPI.

**Table 10.2** Ownership of electric appliances in Khorian and Narian (Percentages)

| Type of appliance | Khorian | Narian |
|---|---|---|
| Any electric appliance | 100 | 95.8 |
| Light bulb or tube light | 100 | 95.8 |
| Television | 0 | 0 |
| Washing machine | 0 | 0 |
| Iron | 0 | 0 |
| Computer | 0 | 0 |
| Ceiling or pedestal fan | 84.6 | 91.3 |
| Air cooler or conditioner | 0 | 0 |

Energy Poverty Survey (2008–2009)—Author's own conducted survey

**Table 10.3** Descriptive statistics of energy source usage and occupation in two RET communities

| | Khorian | Narian |
|---|---|---|
| *Energy sources used (number of households)* | | |
| Firewood (bought and collected) | 26 | 24 |
| Animal waste | 6 | 1 |
| Plant waste | 0 | 0 |
| Kerosene | 26 | 24 |
| Liquid petroleum gas | 0 | 0 |
| *Income (Rs.) and occupation (%) of the households* | | |
| Household average monthly income (Rs.) | 6761.00 | 4035.50 |
| Unemployed (%) | 11.50 | 0.00 |
| Farming (%) | 73.00 | 78.00 |
| Labor (%) | 7.60 | 13.00 |
| Shopkeeper (%) | 3.80 | 4.00 |
| Government employee (%) | 3.80 | 0.00 |
| Private employee (%) | 0.00 | 4.00 |

Energy Poverty Survey (2008–2009)—Author's own conducted survey

mix. Interestingly, households with solar PV systems still incur even more inconveniences than users of traditional energy sources, which include firewood, plant and animal waste, and kerosene. This is reflected in the number of energy sources used by households, which averages three for both categories, although the range is smaller for RET households (3–4) than for non-RET households (1–6).

The Energy Shortfall Index (ESI) indicates the shortfall of energy by comparing actual energy consumption with an adjusted energy threshold[4] of 0.7 metric tons of

[4]According to Pokharel (2006), attaining a Human Development Index (HDI) of 0.8 requires one metric ton of oil equivalent (TOE) per annum per capita in advanced economies. We lowered this value to 0.7 TOE to account for the fact that our survey focused on rural communities in a developing country.

**Table 10.4** Comparison of energy poverty variables between RET and non-RET households

| Variable | RET households | | | | Non-RET households | | | |
|---|---|---|---|---|---|---|---|---|
| | N | Minimum | Maximum | Mean | N | Minimum | Maximum | Mean |
| Number of energy sources used by a given household | 50 | 3.00 | 4.00 | 3.14 | 590 | 1.00 | 6.00 | 3.26 |
| Energy Inconvenience Excess Index (EIE) | 50 | −62.45 | 41.33 | −11.71 | 590 | −156.46 | 100.00 | 25.82 |
| Energy Shortfall Index (ESI) | 50 | −95.39 | −25.05 | −74.98 | 590 | −99.45 | 249.97 | −69.65 |
| Energy Poverty Index (EPI) | 50 | −61.80 | 0.44 | −43.35 | 590 | −92.74 | 90.77 | −21.91 |

Energy Poverty Survey (2008–2009)—Author's own conducted survey oil equivalent (TOE). A positive ESI value indicates no shortfall or energy abundance (per capita), whereas a negative value indicates the degree of energy shortfall. The Energy Poverty Index (EPI) is the non-weighted mean of EIE and ESI and expresses relative energy poverty per capita. RET households show negative mean values of EIE, ESI, and EPI. This implies that even though they have access to RET, they must still be considered energy-poor. Indeed, the mean EIE value for RET households is even clearly worse than for non-RET households, indicating that solar PV systems play an insignificant role in improving RET households' energy poverty situation.

### *10.3.3 Perception of RET by Non-RET Households*

We asked households in non-RET communities whether they would use RET at home and at their workplace (within the community) if such an option were available. Regarding usage at home, nearly 63 % of responded that they did not know yet, whereas nearly 24 % of responded that they would prefer RET to conventional sources of electricity. Responses were similar with regard to using RET at work. Furthermore, we found that over 54 % of the households interviewed had never heard of any technology that could help them access electricity in the absence of a grid connection.

When asked about their preference for different types of RET, including biogas plants, wind energy technology, solar energy technology, and micro-hydro technology, more than 67 % of respondents had no clear preference because they had little or no knowledge about these systems ("Don't know"). More than 27 % of respondents preferred solar energy technology for a variety of reasons. In addition to the "Don't know" response for non-RET households who lacked knowledge about RET, there was also a "No preference" response for those who had knowledge about RET (but no experience of using them) and were so dissatisfied with the available solutions that they would prefer any RET solution in lieu of conventional sources of electricity. However, with just over 2 %, the share of "No preference" responses turned out to be negligible in comparison with the "Don't know" category.

Next, we also asked respondents about the main factors influencing their preference of RET to conventional electricity sources. Possible answers were as follows: convenience, cost, ease of maintenance, technology life span, replacing conventional electricity, and other reasons (including liking, recommendation by others, no preference, no reason). Nearly 60 % of respondents indicated cost as their main reason for switching to RET. Installation of RET would free them from installation charges and monthly utility bills that they currently incurred for grid electricity. Further, more than 10 % of respondents thought that RET are less prone to blackouts, which they viewed as an added advantage. With respect to different types of RET solutions, 22 % of respondents preferred community-based systems; just above 3 % preferred an independent, household-based solution, whereas 61 % had no particular preferences and 14 % did not know.

### 10.3.4 Willingness to Pay for RET

Table 10.5 shows households' willingness to pay for independent RET solutions (one-time payment) and community-based RET solutions (monthly contributions), respectively. In both cases, more than 63 % of the households responded that they were not willing to pay for the proposed solution.

In the case of independent RET solutions, only 4.7 % of respondents were willing to pay 10,000 at the most. Similarly, in the case of community-based RET solutions, a total of only 16.7 % of the households were willing to pay Rs. 100–500 per month.

### 10.3.5 Satisfaction with RET in RET Households

According to the AEDB, each household in Khorian and Narian had been provided with one or more 80-Watt solar panels, a charge controller, a battery, 4 compact fluorescent lamps, 2 light-emitting diode (LED) lights, a 12-Volt direct-current (DC) fan, and a TV socket. We asked local households using a solar PV system whether they think that this technology will be able to meet energy demands in the future. All 50 respondents were of the clear opinion that solar PV systems would not be able to meet future energy demands, neither at the household level nor at their workplaces. To assess the technology's impact on education and health (and, thus, on overall rural development), we asked rural households about their opinion on alternative energy technologies in general and solar PV systems in particular.

**Table 10.5** Willingness to pay

| Independent RET solution | | | Community-based RET solution | | |
|---|---|---|---|---|---|
| Willingness to pay (one-time payment, in Rs.) | Frequency | Percent | Willingness to pay (monthly contribution, in Rs.) | Frequency | Percent |
| Nothing at all | 373 | 63.2 | Nothing at all | 374 | 63.4 |
| <10,000 | 28 | 4.7 | 100–200 | 18 | 3.1 |
| 10,000–50,000 | 8 | 1.4 | 200–300 | 36 | 6.1 |
| 50,000–100,000 | 5 | 0.8 | 300–400 | 26 | 4.4 |
| 100,000–200,000 | 2 | 0.3 | 400–500 | 18 | 3.1 |
| 200,000–500,000 | 2 | 0.3 | 500–1000 | 7 | 1.2 |
| | | | >1000 | 2 | 0.3 |
| Total respondents | 418 | 70.8 | | 481 | 81.5 |
| Missing | 172 | 29.2 | | 109 | 18.5 |
| Total | 590 | 100.0 | | 590 | 100.0 |

Energy Poverty Survey (2008–2009)—Author's own conducted survey

Among other questions, we asked them how satisfactory their present energy situation was after including solar PVs in their energy supply. Once again, all respondents replied that their energy situation was not at all satisfactory.

## 10.4 Reasons for the Failure of Solar PV Projects

Upon discussing several aspects of the solar PV projects' failure with the local residents of Khorian and Narian, we found that most of them saw two major reasons for that failure. These two reasons are discussed separately in the following subsections.

### *10.4.1 Lack of Knowledge About Households' Demand*

The AEDB started the project in Khorian and Narian with an initial study of both rural communities in 2003–2004. After selecting the site, they mandated a UK-based firm, Akhtar Technologies, to supply the solar panels, batteries, LED lights, and other essential materials required for the initial functioning of solar PV cells. The AEDB's implementation strategy completely lacked any analysis of households' energy needs and energy consumption levels, even though research has shown that this is an essential step to be taken prior to technology deployment (Biswas and Lucas 1997). Instead, the AEDB focused more on the suitability of the site: An interview with an employee at the AEDB confirmed that the agency's main concern had been to select sites with sufficient levels of annual sunlight to ensure that enough electricity could be produced to meet domestic electricity needs.

By contrast, our respondents in Khorian and Narian indicated as one of the reasons for their dissatisfaction with the solar PV systems installed that these systems fail to meet households' energy demand. For example, respondents pointed out that a major reason why they did not own high-voltage appliances was that the energy supply generated by their 80-Watt solar panels is insufficient to run such appliances. This was confirmed by the survey team. According to our respondents, the AEDB initially promised rural households that they would periodically upgrade the system in view of local demand, but later, the agency reneged on these assurances. As a result, there remains an enormous gap between the electricity supply generated by the solar PV systems installed and users' demands.

### *10.4.2 Maintenance and Follow-up*

The second major problem that local residents mentioned was that the AEDB and Akhtar Technologies had not conducted any maintenance activities or follow-up

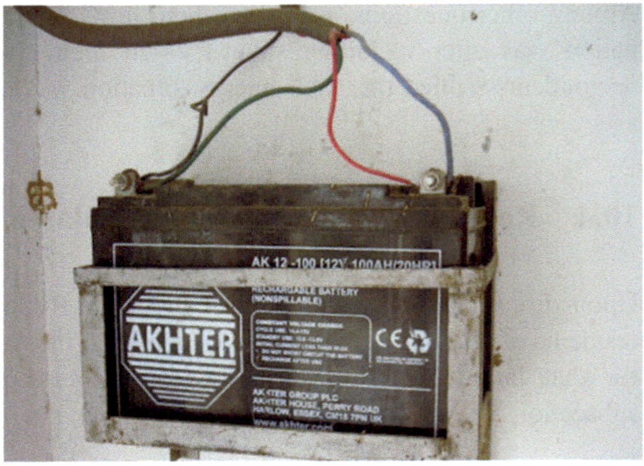


**Fig. 10.2** Rechargeable batteries provided by Akhtar Technologies that were non-functional for several months

visits once the solar panels had been installed (Fig. 10.2). As a result, many solar panels had stopped working at some point. On our visit during the survey, several households showed us dysfunctional charge controllers and LED lamps. The trend of leaving the villagers on their own had increased over time, as mentioned by the local residents of Khorian and Narian. According to one of the respondents, "all these appliances don't have any value for us, as they are hardly used." A few of the respondents acknowledged that the AEDB and Akhtar Technologies had provided initial support during the first few months following the installation. However, as time passed, both the government agency and the private company had ceased to visit the villages for the periodical follow-up maintenance required to keep the solar panels running efficiently.

## 10.5 Conclusions

### *10.5.1 The Causes of Disillusion*

In this paper, we analyze perceptions of RET among households with no access to these technologies as well as the satisfaction of households actually using solar PV systems with these systems and with other RET. Our results reveal a number of major discrepancies between the perceptions of RET in non-RET communities and user households' satisfaction with solar PV systems in RET communities. Recognizing that their energy demands might not be met by conventional energy sources, rural households in non-RET communities had extremely high expectations of RET, especially of solar PV systems. By contrast, households who had already had solar PV systems installed showed high levels of resentment, anger, and disappointment with the technology. The government's implementation strategy had not included any analysis of households' energy needs and energy consumption levels. Accordingly, we found an enormous gap between the electricity supply generated by the solar PV systems installed and users' demands. Moreover, the project completely lacked community involvement and hence failed to build a sense of ownership among local residents. This, together with a lack of maintenance, has caused complete disillusionment among rural households using solar PV systems; they do not believe that solar energy could provide a solution to their energy problems. Our findings in Khorian and Narian further reinforce the research results of Martinot et al. (2001) and Balinet (2006) that merely providing appropriate technology to the poor is not enough to ensure success.

### *10.5.2 Recommendations*

In order to ensure that technology deployment contributes to development and poverty alleviation, we recommend that, before approving such projects, the AEDB performs a substantive analysis of potential beneficiaries' energy needs—in addition to assessing and ensuring other criteria and standards as is currently done. Moreover, based on our findings regarding current perceptions of RET among nonusers and negative user feedback on RET, we expect that government agencies, private companies, international donors, and nongovernmental organizations may face considerable challenges in achieving a sustainable RET-based energy supply with the consensus of local communities; this consensus, however, is essential. Accordingly, local agencies for the promotion of alternative energy technologies need to work toward greater user satisfaction by improving service delivery mechanisms and focusing on the sustainability of RET projects, with a view to scaling them up at a commercial level. This could not only improve energy access, but could also eventually build up capabilities, develop local markets, and promote production capacities for such technologies in the country.

**Acknowledgments** The author would like to express his deep gratitude to Prof. Adam Szirmai and Prof. René Kemp from United Nations University-MERIT, Maastricht University, the Netherlands, for their continued guidance and support in designing and executing the survey, as well as for supervising his doctoral study, which provided the basis for this paper. Moreover, the author would also like to thank the reviewers, particularly Dr. Albrecht Ehrensperger and the participants of the 3rd UNESCO Conference on Technologies for Development (Tech4Dev 2014) in Lausanne, Switzerland, for their helpful comments.

## References

Alam, M., Rahman, A., & Eusuf, M. (2003). Diffusion potential of renewable energy technology for sustainable development: Bangladeshi experience. *Energy for Sustainable Development, 7* (2), 88–96.

Bajracharya, D., Nakarami, A. M., & Singh, K. M. (1991). Development of micro-hydro systems in Nepal: Problems and prospects. ICIMOD Mountain Infrastructure and Technology Series

No. 10 http://lib.icimod.org/record/25364/files/attachment_424.pdf?version=1. Accessed 24 February 2015.

Balinet, P. J. (2006). Bringing solar home systems to rural El-Salvador: Lessons for small NGOs. *Energy Policy, 34*(6), 721–729.

Bhattacharya, S. C., & Jana, C. (2009). Renewable energy in India: Historical developments and prospects. *Energy, 34*(8), 981–991.

Bikam, P., & Mulaudzi, D. J. (2006). Solar energy trial in Folovhodwe South Africa: Lessons for policy and decision-makers. *Renewable Energy, 31*(10), 1561–1571.

Biswas, W. K., & Lucas, N. J. D. (1997). Energy consumption in the domestic sector in a Bangladesh village. *Energy, 22*(8), 771–776.

Biswas, W. K., Bryce, P., & Diesendorf, M. (2001). Model for empowering rural poor through renewable energy technologies in Bangladesh. *Environmental Science and Policy, 4*(6), 333–344.

Chandrasekar, B., & Kandpal, T. C. (2007). An opinion survey based assessment of renewable energy technology development in India. *Renewable and Sustainable Energy Reviews, 11*, 688–701.

de Martino Jannuzzi, G., & de Melo, C. A. (2013). Grid-connected photovoltaic in Brazil: Policies and potential impacts for 2030. *Energy for Sustainable Development, 17*(1), 40–46.

Dendukuri, G., & Mittal, J. P. (1993). Household energy needs of a village in the Rayalaseema area of Andhra Pradesh. *India Energy Conversion and Management, 34*(12), 1273–1286.

Fara, S., Finta, D., & Micu, G. (1998). Problems of village electrification based on PV systems in Romania: Individual solar home systems for settlements in the Cerna Valley. *Renewable Energy, 15*(1–4), 519–522.

Farhar, B. C., & Houston, A. H. (1996). Willingness to pay for electricity from renewable energy. (No. NREL/TP-460-21216): National Renewable Energy Laboratory, US Department of Energy. http://www.nrel.gov/docs/legosti/old/21216.pdf. Accessed 24 February 2015.

Farooq, M. K., & Kumar, S. (2013). An assessment of renewable energy potential for electricity generation in Pakistan. *Renewable and Sustainable Energy Reviews, 20*, 240–254.

Gullberg, M., Ilskog, E., Katyega, M., & Kjellström, B. (2005). Village electrification technologies—an evaluation of photovoltaic cells and compact fluorescent lamps and their applicability in rural villages based on a Tanzanian case study. *Energy Policy, 33*(10), 1287–1298.

Howells, M. I., Alfstad, T., Victor, D. G., Goldstein, G., & Remme, U. (2005). A model of household energy services in a low-income rural African village. *Energy Policy, 33*(14), 1833–1851.

Huraib, F. S., Hasnain, S. M., & Alawaji, S. H. (1996). Lessons learned from solar energy projects in Saudi Arabia. *Renewable Energy, 9*(1–4), 1144–1147.

Kessides, I. N. (2013). Chaos in power: Pakistan's electricity crisis. *Energy Policy, 55*, 271–285.

Lemaire, X. (2011). Off-grid electrification with solar home systems: The experience of a fee-for-service concession in South Africa. *Energy for Sustainable Development, 15*(3), 277–283.

Mala, K., Schläpfer, A., & Pryor, T. (2009). Better or worse? The role of solar photovoltaic (PV) systems in sustainable development: Case studies of remote atoll communities in Kiribati. *Renewable Energy, 34*(2), 358–361.

Martinot, E., Cabraal, A., & Mathur, S. (2001). World Bank/GEF solar home system projects: experiences and lessons learned 1993-2001. *Renewable and Sustainable Energy Reviews, 5*, 39–57.

Millinger, M., Mårlind, T., & Ahlgren, E. O. (2012). Evaluation of Indian rural solar electrification: A case study in Chhattisgarh. *Energy for Sustainable Development, 16*(4), 486–492.

Mirza, B., & Szirmai, A. (2010). Towards a new measurement of energy poverty: A cross-community analysis of rural Pakistan. UNU-MERIT Working Paper Series #2010-024, United Nations University, Maastricht Economic and social Research and training centre on Innovation and Technology.

Obeng, G. Y., Evers, H.-D., Akuffo, F. O., Braimah, I., & Brew-Hammond, A. (2008). Solar photovoltaic electrification and rural energy-poverty in Ghana. *Energy for Sustainable Development, 12*(1), 43–54. doi:10.1016/S0973-0826(08)60418-4.

Palit, D. (2013). Solar energy programs for rural electrification: Experiences and lessons from South Asia. *Energy for Sustainable Development, 17*(3), 270–279.

Platt, L., & Wilson, G. (1999). Technology development and the poor/marginalised: context, intervention and participation. *Technovation, 19*(6–7), 393–402.

Pokharel, G. R. (2006). Promoting sustainable development by creating enterprises on renewable energy technologies in Nepal: Case studies based on micro hydropower projects. Universitat Flensburg. http://www.zhb-flensburg.de/dissert/pokharel/pokharel.pdf. Accessed 24 February 2015.

Rajabapaiah, P., Jayakumar, S., & Reddy, A. K. N. (1993). Biogas electricity: The Pura Village case study. In T. B. Johansson, H. Kelly & A. K. N. Reddy (Eds.), *Renewable Energy*. Washington DC: Island Press.

Thompson, G. A., & Singh, D. (1996). Implementation of village electrification projects in developing countries: The role of CASE. World Renewable Energy Council (WREC) 1996 paper.

Wamukonya, N. (2007). Solar home system electrification as a viable technology option for Africa's development. *Energy Policy, 35*(1), 6–14.

Zein, A.-E.-H., & Sarar, W. A. (1998). Analysis of solar photovoltaic-powered village electrification at Abou-Sorra in Damascus Region. *Renewable Energy, 14*(1–4), 119–128.

# Chapter 11
# Decision-Making and Planning Framework to Improve the Deployment Success of Decentralized Rural Electrification in India

**Abhishek Jain and Paul Kattuman**

**Abstract** 300 million people in India lack access to electricity. More than 90 % of them live in rural areas. The low electrification rate of around 55 % among rural households and the lack of electrification infrastructure in remote areas are major challenges. Decentralized rural electrification has emerged as an alternative approach to electrifying remote rural areas; but an increasing number of commissioned projects are failing to stand the test of time. The present research identifies what factors were crucial in successfully translating off-grid technologies into sustainable solutions. The findings indicate that in addition to appropriate technology adoption, the success of decentralized rural electrification depends on proper management of socioeconomic, operational, environmental, and economic challenges. This includes adequate needs assessment, awareness raising, tariff sensitivity to socioeconomic conditions, demand management, capacity building, access to patient capital, a high capacity utilization level, and sensitivity toward the local environment. The transition from few success stories to scaling up calls for standardizing the planning approach while customizing the solutions. Based on these findings, we developed a multi-tier decision-making framework to assist the planning and management of decentralized rural electrification and enhance the sustainability of projects. The framework systematically captures the complexity of factors that need to be considered for effective decision-making and planning of decentralized rural electrification. We anticipate that implementation of the proposed framework will improve the success rate of projects, enabling decentralized rural electrification to be scaled-up as a sustainable electrification solution.

A. Jain (✉)
Council on Energy, Environment and Water, New Delhi, India
e-mail: abhishek1526@gmail.com; abhishek.jain@ceew.in

P. Kattuman
University of Cambridge, Cambridge, UK



S. Hostettler et al. (eds.), *Sustainable Access to Energy in the Global South*,
DOI 10.1007/978-3-319-20209-9_11

## 11.1 Introduction

Rural electrification plays a critical role in alleviating poverty and facilitating socioeconomic growth (Chakrabarti and Chakrabarti 2002; DfID 2002; Sovacool and Drupady 2012). It is a prerequisite to achieving sustainable development of rural communities (UNDP 2005; World Bank 2008). The impact of energy access, including electrification, on progress toward the Millennium Development Goals has been widely discussed (Barkat et al. 2002; Bastakoti 2006; United Nations Development Programme [UNDP] 2005).

But even today, 1.3 billion people across the globe are yet to realize these benefits (IEA 2012). The vast majority of these people live in two main regions: sub-Saharan Africa (590 million) and developing Asia (628 million). At the country level, India has the largest population (about 300 million) without access to electricity (IEA 2012). More than 90 % of this unelectrified population lives in rural areas.

### *11.1.1 Rural Electrification in India*

Rural electrification in India has been accomplished predominantly by extending the centralized grid to rural areas. But the grid has not been able to provide electricity to rural communities in a reliable manner. Many rural areas lack electricity in peak hours when it is needed the most (Palit and Chaurey 2011). Such load-shedding is mainly due to the grid having been extended without simultaneously expanding generation capacity, as well as the prioritization of urban and industrial consumers over rural consumers (Kemmler 2006; Krishnaswamy 2010). The unreliable electricity supply, inadequate pricing mechanisms, and the fact that some rural households cannot afford an electricity connection are the main reasons why the electrification rate among rural households in India is still only about 55 %, even though the grid has reached more than 97 % of villages. In light of these challenges, decentralized electrification provides a technically sound, locally manageable, and cost-effective alternative to grid extension.

Decentralized electrification essentially means local generation and distribution of electricity, eliminating the need for transmission. The generation technologies, mostly based on renewable resources, include solar photovoltaic (PV) systems, micro-hydro systems, wind turbines, and biomass gasification. As a broader concept, it includes solutions ranging from battery boxes or solar lanterns to individual household systems and village-level micro-grids. The present research considers this broad purview of decentralized electrification while keeping a particular focus on village-level micro-grids.

### 11.1.2 Decentralized Rural Electrification (DRE) in India

Most off-grid rural electrification projects are accomplished under the remote village electrification (RVE) program of the Ministry of New and Renewable Energy (MNRE). As of June 2013, the MNRE had electrified over 9000 villages and hamlets with a mixture of biomass gasifiers ($\sim$17 MW), solar PV systems ($\sim$132 MW), and more than 2000 micro-hydro projects. In addition to these government initiatives, the private sector—including social entrepreneurs and nongovernmental organizations (NGOs)—has taken significant steps toward decentralized rural electrification (DRE).

Despite the progress, DRE in India has yielded mixed results. Many micro-grids, solar household systems, and solar lanterns are defunct (Palit and Hazarika 2002); programs for community-based micro-hydro projects in Leh region have partially failed (Ete and Prochaska 2009); and the MNRE's Village Energy Security Programme (VESP) has been discontinued due to a high rate of dysfunctional systems (Palit et al. 2013).

### 11.1.3 Existing Research on the Limited Success of DRE

Recent literature points to a number of socioeconomic, technical, and institutional issues that need to be resolved in order to achieve long-term sustainability of DRE projects (Gambhir et al. 2012). A large number of DRE projects have failed because they focused mainly on technological aspects of installation and gave insufficient attention to long-term project sustainability, which depends on multiple interlinked factors spanning technical, financial, regulatory, and institutional areas (Kumar et al. 2009; Palit 2003).

Various case studies of pilot projects, policy appraisals, and advocacy studies have also increased the understanding of the sector. Bhattacharyya (2012) provides a comprehensive review of approaches and techniques used in the past. However, very few attempts have been made to use past learnings in a structured and systematic manner to improve planning and decision-making for future projects and enhance their sustainability and success. The present work fills this gap by proposing a multi-tier decision-making framework for planning of future DRE projects.

## 11.2 Research Approach and Methodology

Three sequential methodological approaches were used in this study: (1) qualitative analysis of existing case studies using a general inductive approach, (2) unstructured interviews with multiple stakeholders involved in DRE, and (3) site visits to two micro-grid installations for validation of emerging research patterns against ground realities.

### 11.2.1 Qualitative Case Study Analysis

A case-study-based approach was used to comprehend the dynamics of factors affecting the design, implementation, and long-term sustainability of DRE projects. Six case studies were selected from existing literature so as to (1) incorporate diverse contexts, (2) achieve a balanced mix of successful and failed cases, and (3) include case studies in different phases of scale-up. We adopted a general inductive approach (Thomas 2006)—a qualitative assessment methodology—to gain a comprehensive understanding of the issues.

The six case studies analyzed are listed in Table 11.1. Four studies focus on India and two on other countries experiencing electrification access challenges. Looking at more than one country provides a greater understanding of the kind of issues faced and solutions conceived by proponents to tackle them. The aim of the analysis was to capture common patterns as well as divergent, context-specific aspects. The diversity of case studies in terms of geographical location, technologies employed, and operational and financial models adopted enabled us to identify strong common themes and subtle connections between different underlying factors.

One important observation made during the collation of case studies was that very limited documentation exists for the failed cases of micro-grid projects in India. Interviews with stakeholders and field visits suggested that the number of failed cases is much greater than what is reported or published, indicating the existence of a selection bias.

### 11.2.2 Unstructured Interviews with Stakeholders and Field Visits

In order to strengthen the findings from case study analysis, we conducted unstructured interviews to draw from the experience, opinions, and concerns of multiple stakeholders involved in DRE in different capacities. These included private developers, policy advocates, consultants, and academic researchers. The interviews were conducted on the condition of anonymity. We chose an unstructured interview approach to avoid potential biases or constraints introduced by the design and selection of interviewer's questions, as in case of structured or semi-structures interviews. Table 11.2 lists the stakeholders interviewed. Government officials involved in rural and decentralized electrification were not available for interviews. In order to understand policymakers' views, we conducted a qualitative assessment of existing policy documents and recent policy proposals.

In addition to the interviews with stakeholders, we conducted field visits to the DRE sites. They served two purposes: (1) to obtain first-hand information on the perspective of the most important stakeholder group—final consumers and beneficiaries; and (2) to validate findings and outcomes of the analysis against ground realities.

**Table 11.1** Overview of case studies analyzed

| S. No. | Program/project/organization | Country | Technology | Agencies involved | Type of organization/delivery model | Financial model | Coverage | Status | Timeline | Sources |
|---|---|---|---|---|---|---|---|---|---|---|
| 1 | Village Energy Security Programme (VESP) | India | Biomass gasification, biogas-/biofuel-run generators, micro-grids | MNRE; World Bank | MNRE initiative; village energy committee leads design, implementation, and operation, supported by state agencies | 90 % state-funded, 10 % Community contribution | 79 projects sanctioned across 9 states; 65 projects commissioned (700 kW of cumulative capacity) | Less than 50 % of projects are fully or partially functional | 2004–2012; stopped due to limited success | Palit et al. 2013; interviews |
| 2 | Ladakh Ecological Development Group (LEDeG) | India | Micro-hydro systems | BORDA; European Union; MNRE | NGO-facilitated development and management, supported by international promoters; operated and maintained by community elected committee | 50 % Donor grants, 50 % community contribution (cash and kind); operation and maintenance sustained by monthly revenues | More than 70 community-based micro-hydro projects; capacity ranging between 0.5 and 30 kW | Almost 25 % of projects are not operational | 1989–ongoing | Ete and Prochaska 2009; Pareek and Ladakh Ecological Development Group 2007 |
| 3 | Grameen Shakti (GS) | Bangladesh | Solar household systems (SHS) | IFC; USAID; Government of Bangladesh | Self-sustainable, nonprofit enterprise; installs and maintains systems | More than 90 % revenues from sales only; few tax exemptions; initially supported by grants and loans | 1.3 million SHSs installed; system size between 10 and 130 Wp, typical size 50 Wp (Wp indicates rated capacity of solar panels) | Fairly successful, phenomenal expansion, but leaves out the very poor | 1996–ongoing | Grameen Shakti 2012; Sovacool and Drupady 2011 |
| 4 | Mera Gao Power (MGP) | India | Solar-PV-based direct current (DC) micro-grid | USAID | For-profit enterprise; installs, owns, operates, and maintains the micro-grid | Largely self-sustaining through community revenues; initially supported by grants | 100 villages/hamlets; typical system size of 0.5 kWp; ~5 W connection per household (2 LED lamps and mobile charging) | Unique model fitting very basic needs; challenges in scaling up the system beyond a certain limit, as the aspiration level increases. | 2010–ongoing | Ballesteros et al. 2012; interview |

(continued)

**Table 11.1** (continued)

| S. No. | Program/project/organization | Country | Technology | Agencies involved | Type of organization/delivery model | Financial model | Coverage | Status | Timeline | Sources |
|---|---|---|---|---|---|---|---|---|---|---|
| 5 | Rural Energy Development Program (REDP) | Nepal | Micro-hydro systems | Government of Nepal; UNDP; World Bank | Participatory program led by Government of Nepal and supported by UNDP; implemented, operated, and maintained by district and village committees | 50 % donors and government; 40 % community; 10 % village and district committees | Cumulative installed capacity of 5500 kW benefitting around 50,000 households across 40 districts | Successful continuation and scaling of micro-hydro in Nepal, post-exit of donors and promoters | 1996–2011 | Sovacool and Drupady 2012; UNDP-APRC 2012 |
| 6 | Husk Power System (HPS) | India | Biomass gasification with pure producer gas engine | IFC; Shell Foundation; MNRE | For-profit enterprise; various business models with mixed combinations of build, own, operate, and maintain | Self-sustaining; subsidy on equipment by MNRE; initially supported by grants | 91 installations reaching 40,000 households; installation size between 25 and 120 kW | Phenomenal expansion and scale-up, but some challenges | 2008–ongoing | Mehta and Mehta 2011; SEVEA 2013; interviews |

*Note* Due to the dynamic nature of certain data and the diversity of sources, the figures do not refer to a unique point of time. Due efforts were made to capture the latest reported numbers. Agency acronyms: MNRE = Ministry of New and Renewable Energy Resources; BORDA = Bremen Overseas Research and Development Association; IFC = International Finance Corporation; USAID = United States Agency for International Development; UNDP = United Nations Development Programme

**Table 11.2** List of stakeholders interviewed

| No. | Stakeholder's role | Experience in renewable-energy off-grid projects (years) |
|---|---|---|
| 1 | Entrepreneur: private developer and operator of DRE projects | 11 |
| 2 | Entrepreneur: private developer and operator of DRE projects | 6 |
| 3 | Entrepreneur: private developer and operator of DRE projects | 5 |
| 4 | Consultant for off-grid electrification projects | 8 |
| 5 | Policy researcher working on electrification access, renewable energy, and electricity sector reforms, for a nonprofit organization | 14 |
| 6 | Academic researcher focusing on DRE, renewable energy, and electrification reforms | 12 |
| 7 | Academic researcher focusing on DRE in South Asia and Africa | 16 |
| 8 | Academic researcher focusing on DRE and micro-grids | 7 |

### *11.2.3 Framework Development*

After a comprehensive analysis of various decision-making approaches regarding their relative advantages and disadvantages for DRE projects (Bhattacharyya 2012), we opted for a multi-tier framework approach. We did so for three reasons:

1. To avoid masking the complexity and interdependency of multiple factors and decisions influencing project sustainability (a risk inherent in standard optimization or modeling-based approaches).
2. To support the iterative nature of decision-making that is required for effective DRE project planning.
3. To maintain the necessary flexibility for DRE planning (missing in rigid approaches such as multi-criteria decision analysis).

## 11.3 Research Findings

One of our main findings is the identification of major challenges faced by DRE projects and the corresponding management strategies employed. These are described and categorized in the following sections. The list is not exhaustive, and not all projects face the same challenges; but the list captures most of the critical issues that need to be addressed to achieve successful, sustainable, and scalable DRE projects. After listing the different challenges, we identify the management levels at which the various challenges are best addressed.

## 11.3.1 Technical and Operational Challenges

**Demand-Side Management**. Various DRE projects—even ones designed on the basis of detailed need assessments—face the challenge of unplanned variability in electricity demand. Most projects do not have actual consumption-based tariffs but charge flat rates with predetermined usage. In the absence of meters, households tend to have erratic usage patterns, at times overshooting the system limits and leading to instability of the micro-grid. Oversizing is the usual but inadequate approach to resolving such situations.

Some developers have opted for technical solutions, attempting to manage demand by installing small-scale limiting devices such as the "electricity dispenser" used by Trama TecnoAmbiental (TTA), a Spanish organization that has been working in DRE for 20 years (Rolland and Giana 2011). However, the added costs of such equipment increase the project's capital cost.

Another challenge that arises in privately managed micro-grids is electricity theft. Entrepreneurs either rely on communal arrangements to check such unwarranted behaviors, or resort to technological solutions such as electricity distribution based on a smart grid with anti-theft mechanisms. Gram Power, a for-profit enterprise working in the state of Rajasthan, develops and operates such smart micro-grids.

**Lack of Human Capacity**. Lack of skilled human resources in remote areas to manage operations and maintenance of off-grid systems, along with insufficient training and limited retention of trained personnel, is a major operational challenge. VESP and Ladakh Ecological Development Group (LEDeG), two programs with various nonfunctional projects, struggled with this challenge. The issue was also repeatedly highlighted by entrepreneurs during interviews.

**Lack of Technical Standardization**. Assessments and field visits revealed that technical standardizations of off-grid projects are nonexistent in India. Cost-cutting measures and limited technical experience lead to the use of substandard material and to inadequate consideration of quality; this, in turn, results in poor performance. Moreover, independent rural off-grid electricity generators do not need a generation license and thus are exempt from adherence to technical standards. Hence, the challenge is not only about a lack of technical standards but also about inadequate regulation.

## 11.3.2 Economic Challenges

**High Capital Investment**. Renewable-energy micro-grid projects are capital intensive. To achieve financial viability, tariffs must be considerably higher than grid electricity tariffs. Moreover, as these projects are in areas where high tariffs surpass consumers' ability to pay, they often experience poor returns. For this reason, private investment into the DRE sector has been limited. Micro-financing emerged as a solution in certain cases, such as Gram Shakti in Bangladesh.

**Lack of Financing**. Raising debt or finance for DRE projects is a major hurdle. An entrepreneur developing micro-grids based on solar PV mentioned that "as banks do not have sufficient validated success cases, and due to huge uncertainty involved in the sound payback of such projects, raising loan from the banks is very difficult." Even though MNRE provides up to 30 % subsidy against initial capital investment, many developers chose not to apply for subsidies because they involve bureaucracy, corruption, as well as delays and uncertainty in payouts.

**Competition with Kerosene**. Most small-scale DRE projects provide only basic lighting and thus have to compete with kerosene—the prevailing lighting alternative—which is also used as a cooking fuel in India and is heavily subsidized. The improvement in luminance and indoor air quality as a result of switching to a cleaner source of light has positive health impacts (Pokhrel et al. 2010), but local communities are not necessarily aware of this.

**Poor Capacity Utilization Factor (CUF)**. CUF indicates the actual energy generation of a plant in a given time compared to the maximum energy that plant can generate in the same time. A low CUF and a low plant load factor (PLF) lead to unreliable operations, as a very low PLF poses technical difficulties. Moreover, low CUF and PLF increase per-unit generation cost and significantly reduce returns on capital investment. Low CUF levels are more evident with technologies such as biomass gasification and micro-hydro, mainly due to limitations in downsizing these technologies or poor assessment of electricity needs. Adequate approaches include using technologies suitable for smaller loads; adding productive loads to the system; or planning projects (for suitable technologies such as solar PV) stage by stage, adding generation capacity as demand increases.

**High Maintenance Cost**. Most DRE projects are in remote, poorly accessible areas where local supply chains for spare parts are nonexistent. This leads to either high maintenance inventory costs or extended breakdown periods that undermine community confidence and hence returns. Unavailability of local competence leads to travel of skilled personnel, thus increasing maintenance cost.

## *11.3.3 Social and Socioeconomic Challenges*

**Inadequate Need Assessment**. Inadequate understanding of consumer needs leads to inappropriate solutions. The type and scale of a service, the pricing model, and various other design aspects must correspond to community needs. As one DRE project developer mentioned, "Understanding that the charging of mobile phones is an equally urgent need as lighting, and hence developing a solution catering to both these needs, resulted in a far greater subscription [to the system]."

A related aspect, raised by another entrepreneur, is community aspirations, which are sometimes much higher than what an off-grid project can cater in an economically viable manner. Thus, engagement, consultation and awareness raising are very important, and need assessment should be understood not as a one-way flow of information from the community toward the planner, but as an engaging process.

**Lack of Awareness**. Lack of awareness about electricity, especially about its productive usage, poses challenges to its adoption by communities. It has also been observed that at times communities lack confidence in DRE and show apprehensions that off-grid electrification will negatively impact on the extension of the centralized grid to the village. Awareness generation is a prerequisite to creating sufficient and sustained demand of electricity, overcoming apprehensions, and ensuring efficient usage.

**Lack of Community Engagement**. Partly associated with the previous two challenges, lack of community engagement critically hinders a project's long-term success. Community engagement includes a collaborative approach to need assessment, transparent information dissemination, and awareness raising. It helps to achieve continuous support from the community right from the start in the form of contributions in project implementation and operation. It also helps to create a sense of ownership of the system among the community. Limited or unsuccessful attempts to engage the community result in partial or complete failure of projects, as observed in some of the case studies analyzed. The importance of community engagement was also validated through discussions with local community members during field visits.

**Inability to Pay**. Consumers' limited ability to pay for electricity is a major challenge. For example, in the state of Chhattisgarh, government initiatives have led to phenomenal coverage of household electrification. But even the operations of DRE projects are subsidized, as consumers' ability to pay does not cover maintenance and battery replacement costs. Cross-subsidizing off-grid consumers is one possible way ahead until rural development leads to improved socioeconomic conditions and greater ability to pay. Inability to pay is also a problem in privately managed projects; in various projects, certain segments of the community could not afford to subscribe to the system. Thus, achieving universal access becomes a financial challenge.

This indicates that basic infrastructure development cannot be considered in isolation from overall development plans. DRE projects which also focused on improving local income and enhancing wider community development by promoting productive electricity use, supporting local enterprises, and imparting training to enhance locals' skills and employability have had more success and have stood the test of time.

## *11.3.4 Policy and Institutional Challenges*

**Future of DRE**. Uncertainties regarding the future of micro-grids limit the private investors' confidence, especially if the centralized grid reaches the location. Building technical feasibility for connecting DRE systems to the grid and developing a distributed generation and utilization network is a possible way forward but requires policy support.

**Overemphasis on Expansion of Centralized Grid**. Policies continue to emphasize grid extension to remote villages, even though household electrification rates and electricity supply have remained low in grid-connected rural areas. Almost 98 % of the rural electrification infrastructure budget under Rajiv Gandhi Gram Vidyutikaran Yojana (RGGVY)—a flagship rural electrification program of the Government of India—is allocated to centralized grid extension. In fact, even 5 years after the rollout of the decentralized distributed generation (DDG) policy in 2009, only one state has commissioned DDG projects, and less than 5 % of the allocated budget has been released (REC India 2014).[1]

**Perception of Decentralized Electrification**. Our assessment of policy documents on decentralized electrification shows that it is perceived as an ad hoc rather than a sustainable solution to electrification. Consequently, there is a lack of policy support for promoting the active participation of public sector utilities and private players in DRE. The current institutional structure and clauses under DDG guidelines discourage private developers and entrepreneurs from bidding for projects under the scheme.

**Unclear Responsibilities**. Multiple institutions are responsible for rural electrification at the central and state levels. Overlapping responsibilities, a lack of collaboration, and a poor flow of information between these institutions have even led to duplication of electricity infrastructure in certain rural areas (Gambhir et al. 2012). This creates confusion and conflict between different agencies and poses challenges for private developers seeking to obtain information about plans for the development and electrification of unelectrified areas.

### *11.3.5 Environmental Challenges*

As DRE technologies are based on renewable resources and do not contribute to global environmental challenges such as climate change, developers generally perceive DRE projects to be inherently environment-friendly. But if renewable-energy micro-grids are to be successful in the long run, it is important that local environmental challenges are adequately managed. This includes issues such as ensuring minimum ecological flows in streams used for micro-hydro projects, sustainable use of biomass for gasification, or responsible recycling of used batteries. There is no policy or regulation in place for end-of-life management of solar PV panels at the national level.

Environmental challenges related to DRE are repeatedly mentioned in the literature and by researchers during interviews. But none of the interviewed

[1]Under the XI Plan (2007–2012), the Government of India allocated 265 billion INR (4.4 billion USD) to RGGVY. Out of this, only 5.4 billion INR (90 million USD) were allocated to a DRE scheme called DDG. Similarly, under the XII and XIII plans, 354 billion INR were allocated to RGGVY, including only 9 billion INR for DDG.

developers considered environmental challenges an important issue hindering the sustainability of projects. This perception was also evident during field visits.

## *11.3.6 Management Level of Challenges*

Table 11.3 indicates major challenges and their respective management level. Some can be managed at project level through sound planning, decision-making, and implementation, while others require interventions at the policy level and depend on the overall ecosystem. Certain challenges call for better management at both levels. The need for high capital investment is the only challenge which is not entirely under the control of project planners or policymakers; it is largely determined by the nature of the technology and its cost.

Although many of these challenges are significant and specific to DRE, some of them are also evident in other development activities and basic service provisions in rural areas, for example, related to clean drinking water, sanitation, and health care. This emphasizes the need to understand and appreciate the importance of factors such as community engagement, transparency, awareness raising, capacity building, affordability considerations, and fiscal support in basic infrastructure development.

**Table 11.3** Key challenges and their management level

| No. | Challenge | Project level | Policy level |
|---|---|---|---|
| 1 | Demand-side management | ✓ | |
| 2 | Lack of human capacity | ✓ | ✓ |
| 3 | Lack of technical standardization | ✓ | ✓ |
| 4 | High capital investment | – | |
| 5 | Lack of financing | | ✓ |
| 6 | Competition with kerosene | ✓ | ✓ |
| 7 | Poor capacity utilization factor (CUF) | ✓ | |
| 8 | High maintenance cost | ✓ | |
| 10 | Inadequate need assessment | ✓ | |
| 11 | Lack of awareness | ✓ | |
| 12 | Lack of community engagement | ✓ | |
| 13 | Inability to pay | ✓ | ✓ |
| 14 | Future of DRE | ✓ | ✓ |
| 15 | Overemphasis on expansion of centralized grid | | ✓ |
| 16 | Perception of decentralized electrification | | ✓ |
| 17 | Unclear responsibilities | | ✓ |
| 18 | Management of local environment | ✓ | ✓ |

## 11.4 Planning and Decision-Making Framework for Decentralized Rural Electrification Projects

Based on the above research findings, we propose a planning and decision-making framework to enhance the sustainability and scalability of DRE. The multi-tier framework (Fig. 11.1) has five components, where first four, (1) key principles, (2) key determinants, (3) major strategic decisions, and (4) sustainability aspects, deal with the planning of individual projects, whereas (5) main scaling factors concern the overall ecosystem in which individual projects are conducted.

The various arrows in Fig. 11.1 represent each element's influence on others. Each sustainability aspect is affected by strategic decisions, which in turn are influenced by key determinants. The dashed arrows under "major strategic decisions" indicate how one decision affects others, while also highlighting the iterative nature of decision-making that is crucial for successful and effective planning of DRE. At a broader level, the framework highlights the complexity of interactions that need to be understood and considered during planning.

### *11.4.1 Key Components of the Framework*

**Key Principles**. These are important values which must be imbibed and adhered to by project proponents. They should become the core foundation upon which DRE projects are planned, implemented, and operationalized. Key principles are important irrespective of the location or context, though the approach to realizing them in practice will strongly depend on the project's context.

Key principles include aspects such as community engagement and transparency, which are essential for enhancing projects' social sustainability and acceptability. Another key principle is to be flexible in developing solutions, so that they can be adapted to match the community's needs and preferences. Multiple case studies showed that this has a significant positive impact on project success. Similarly, capacity building at the institutional, community, and individual levels are important for both sustaining and scaling up projects in the long run.

**Key Determinants**. These are location-specific physical, social, economic, environmental, and cultural characteristics which influence major project planning decisions. Physical aspects include things such as site accessibility and the number and distribution of households. An assessment of the potentials of various resources such as solar energy, wind, and biomass is another important determinant. Sociocultural characteristics capture elements such as community cohesion and identifying decision-makers in the community who can influence important decisions. Assessments of electricity needs and productive demand potential are determinants that have direct and strong implications for major project decisions, for example, on the scale and size of the service. Effective identification of these determinants depends on the level of community engagement. The final critical

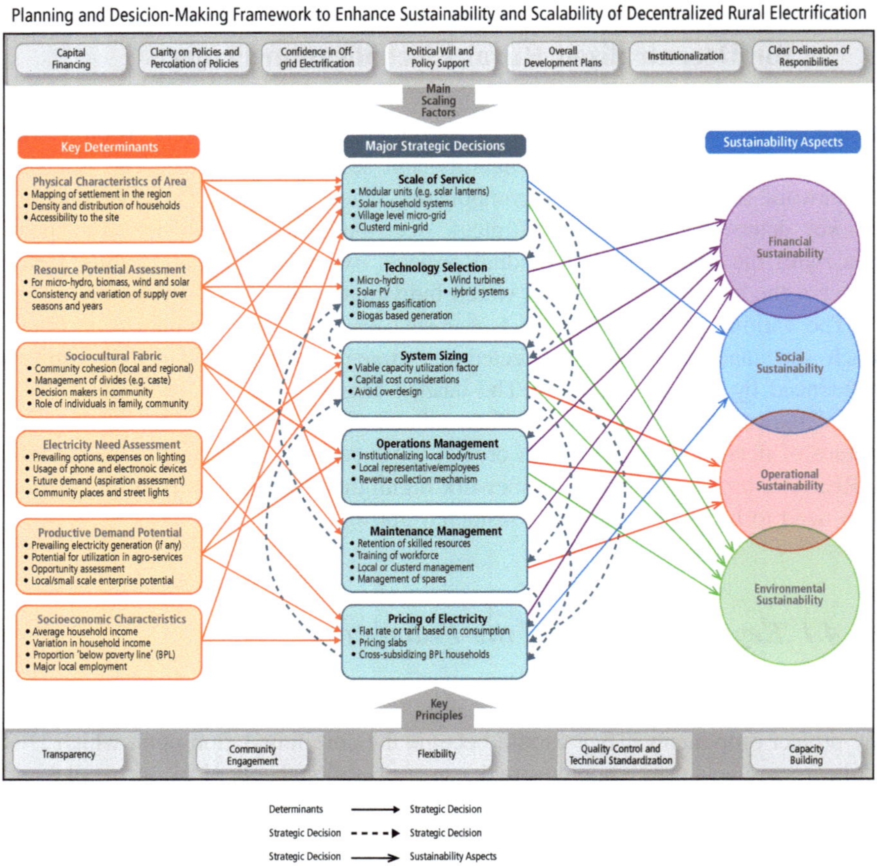


**Fig. 11.1** Project planning and decision-making framework for decentralized rural electrification

determinant is related to socioeconomic characteristics, which directly impact tariff setting and the pricing structure.

**Major Strategic Decisions**. These are high-level planning decisions that dictate the project's basic design and directly affect long-term project sustainability. They include decisions on technical, operational, managerial, and financial aspects. These are directly influenced by key determinants, whereas the key principles provide the environment in which the decisions should be made. Major strategic decisions concern things such as defining the scale of the service, which may range from a simple solar lantern to mini-grids that serve multiple villages; choosing the technology; system sizing; choosing an operational and maintenance model; and pricing.

**Sustainability Aspects**. These are various dimensions of sustainability that determine the long-term success of DRE projects. In the context of DRE, it is appropriate to classify sustainability into social, financial, operational, and environmental dimensions. It is important to note that social sustainability influences all other dimensions of sustainability. Social acceptability and continuous community support are crucial.

**Main Scaling Factors**. These are broader elements within larger sociopolitical and economic systems, which create an enabling environment for scaling up decentralized electrification. Scaling factors include policy support and political will toward DRE and toward the overall development of remote rural areas to create a conducive ecosystem for scaling up DRE. Another critical scaling factor is support through capital financing. DRE is public infrastructure development and hence requires patient capital: Investment should not focus solely on short-term returns. Reducing uncertainties and providing clear policies especially regarding the future of off-grid installations are other important steps toward building trust among DRE proponents. Lastly, institutionalization with clearly defined responsibilities is important in order to streamline efforts toward DRE.

### *11.4.2 Discussion of the Framework*

One major objective of the framework is to combine the learnings from past projects in a useful, concise, and yet comprehensive form that can assist planning and decision-making for future projects. Prevailing planning procedures of state implementation agencies follows a linear approach, often resulting in project failure at later stages. The proposed framework aims to improve decision-making by moving from a linear approach driven by a single denominator to a more comprehensive and balanced approach.

The framework can also serve as a checklist for avoiding common pitfalls that are usually ignored by planners with limited DRE experience. An in-depth discussion on every subcomponent and interaction in the framework cannot be provided here. But additional details and understanding of the interactions between the various components of the framework may be obtained from a report by the author (Jain 2013).

### *11.4.3 Conclusions and Discussion*

Our research findings indicate that DRE can be a sustainable solution which can foster rural electrification in India and other developing countries lacking electricity access. However, it needs to be appropriately planned and implemented. DRE has limitations—such as limited capacity and hours of supply—which can be resolved by scaling up the system as demand increases. This also calls for technical and financial innovations that enable such a system scale-up. Understanding this aspect of DRE is important for changing the prevailing perception of DRE as an ad hoc solution into a view of DRE as a sustainable solution catering to contemporary community needs.

This work also highlights that there is no one-size-fits-all solution. Each DRE solution has to be customized to best fit the context and conditions in which it is implemented, even though the planning approach leading to such solutions has common elements. Hence, scaling up of DRE requires a standardization of planning and decision-making approaches. The long-term success and sustainability of DRE depends on technical, social, economic, environmental, institutional, policy, and regulatory issues. Most of these can be managed by means of effective decision-making at the planning stage and later during implementation and operations. Our proposed multi-tier planning framework incorporates these factors while accommodating the complexity involved in effective decision-making. Effective use of the framework during planning, implementation, and management of future DRE projects can improve their success in the long run.

A higher success rate is important at the individual project level, as this ensures successful utilization of limited resources in populous countries such as India. At the policy and ecosystem level, the cumulative impact of a greater number of successful projects will bolster the confidence of policymakers, investors, and communities in DRE as a sustainable electrification approach for remote and rural areas and foster large-scale deployment of DRE.

## References


Ballesteros, A., Norford, E., Nagle, T., Yonavjak, L., & Alzner, S. (2012). *Implementation strategies for renewable energy services in low-income, rural areas*. Washington, D.C.: World Resource Institute.

Barkat, A., Khan, S., Rahman, M., & Zaman, S. (2002). *Economic and social impact evaluation study of the rural electrification program in Bangladesh*. Dhaka: Human Development Research Centre.

Bastakoti, B. P. (2006). The electricity-livelihood nexus: Some highlights from the Andhikhola Hydroelectric and Rural Electrification Centre (AHREC). Energy for *Sustainable Development, 10*(3), 26–35. doi:10.1016/S0973-0826(08)60541-4

Bhattacharyya, S. C. (2012). Review of alternative methodologies for analysing off-grid electricity supply. *Renewable and Sustainable Energy Reviews, 16*(1), 677–694. doi:10.1016/j.rser.2011.08.033

Chakrabarti, S., & Chakrabarti, S. (2002). Rural electrification programme with solar energy in remote region—a case study in an island. *Energy Policy, 30*(1), 33–42. doi:10.1016/S0301-4215(01)00057-X

Department for International Development [DfID]. (2002). *Energy for the poor: Underpinning the millennium development goals*. London: DFID.

Ete, M., & Prochaska, F. (2009). Determinants of success and failure of community based micro hydro project. In *Proceedings of the Seminar—Energy and Climate Change in Cold Regions of Asia*.

Gambhir, A., Toro, V., & Ganapathy, M. (2012). *Decentralised renewable energy (DRE) micro-grids in India—a review of recent literatue*. Pune: Prayas Energy Group.

Grameen Shakti. (2012). Grameen Shakti makes a world record in non-grid power generation. http://gshakti.org/index.php?option=com_content&view=category&layout=blog&id=54&Itemid=78. Accessed 3 March 2014.

International Energy Agency [IEA]. (2012). *World energy outlook.* Paris: IEA.
Jain, A. (2013). *Development of decision analysis framework to assist decentralized electrification in rural India.* Cambridge: University of Cambridge.
Kemmler, A. (2006). Regional disparities in electrification of India—do geographic factors matter? Zurich: Centre for Energy Policy and Economics (CEPE).
Krishnaswamy, S. (2010). *Shifting of goal posts—rural electrification in India: A progress report.* Bangalore: Vasudha Foundation.
Kumar, A., Mohanty, P., Palit, D., & Chaurey, A. (2009). Approach for standardization of off-grid electrification projects. *Renewable and Sustainable Energy Reviews, 13*(8), 1946–1956. doi:10.1016/j.rser.2009.03.008
Mehta, C., & Mehta, K. (2011). A design space and business strategy exploration tool for infrastructure-based ventures in developing communities. *International Journal for Service Learning in Engineering, 6*(2), 30–57.
Palit, D. (2003). Renewable energy in Northeast India: Issues and prospects. In P. Upender & M. P. Poonia (Eds.), *Energy technologies for sustainable development* (pp. 85–93). Ghaziabad: Prime Publishing House.
Palit, D., & Chaurey, A. (2011). Off-grid rural electrification experiences from South Asia: Status and best practices. *Energy for Sustainable Development, 15*(3), 266–276. doi:10.1016/j.esd.2011.07.004
Palit, D., & Hazarika, N. (2002). Performance and impact of solar photovoltaic devices disseminated for village electrification in Northeast India. In A. K. M. Sadrul Islam & D. G. Infield (Eds.), *Proceedings of the International Conference on Renewable Energy for Rural Development, Bangladesh University of Engineering and Technology, Dhaka* (pp. 13–18).
Palit, D., Sovacool, B. K., Cooper, C., Zoppo, D., Eidsness, J., Crafton, M., et al. (2013). The trials and tribulations of the Village Energy Security Programme (VESP) in India. *Energy Policy, 57,* 407–417. doi:10.1016/j.enpol.2013.02.006
Pareek, L., & Ladakh Ecological Development Group. (2007). *A study on micro hydro units in Ladakh to analyse their feasibility using quantitative & qualitative tools, institutional arrangements, and costing methods* (First). Dehra Dun: Bishen Singh Mahendra Pal Singh.
Pokhrel, A. K., Bates, M. N., Verma, S. C., Joshi, H. S., Sreeramareddy, C. T., & Smith, K. R. (2010). Tuberculosis and indoor biomass and kerosene use in Nepal: A case-control study. *Environmental Health Perspectives, 118*(4), 558–564. doi:10.1289/ehp.0901032
REC India. (2014). Status of sanctioned DDG projects. *Rural Electrification Corporation.* http://www.recindia.nic.in/download/DDG_MIS%2031.01.2014%20.pdf. Accessed 4 March 2014.
Rolland, S., & Giana, G. (2011). *Hybrid mini-grids for rural electrification: Lessons learned.* Brussels: Alliance for Rural Electrification.
SEVEA. (2013). *Case study: Husk power system—power to empower.* Les marches, France: Synergie pour l'Echange et la Valorisation des Entrepreneurs d'Avenir.
Sovacool, B. K., & Drupady, I. M. (2011). Summoning earth and fire: The energy development implications of Grameen Shakti (GS) in Bangladesh. *Energy, 36*(7), 4445–4459. doi:10.1016/j.energy.2011.03.077
Sovacool, B. K., & Drupady, I. M. (2012). *Energy access, poverty, and development: The Governance of small-scale renewable energy in developing Asia* (p. 306). New York: Ashgate.
Thomas, D. R. (2006). A general inductive approach for analyzing qualitative evaluation data. *American Journal of Evaluation, 27*(2), 237–246. doi:10.1177/1098214005283748
United Nations Development Programme [UNDP]. (2005). *Energizing the millennium development goals. A guide to energy's role in reducing poverty.* New York: UNDP.
UNDP-APRC. (2012). Energy to move rural Nepal out of poverty: The rural energy development programme model in Nepal. Bangkok: United Nations Development Programme.
World Bank. (2008). *Designing sustainable off-grid rural electrification projects: Principles and practices.* Washington: The World Bank.

# Part IV
# Potential of Renewable Energy Technologies

# Chapter 12
# Up-Scaling and Mainstreaming Renewable Energy Technologies for Energy Security, Climate Change, and Economic Development

**Pankaj Agarwal and Kinsuk Mitra**

**Abstract** World's energy demand relies heavily on carbon-based fossil fuels which are not only finite but have also been shown to have a negative long-term impact on climate. As world demand for energy rises, renewable energy (RE) has a significant potential to mitigate global climate change, address regional and local environmental concerns, reduce poverty, and increase energy security. The main focus toward RE mainstreaming has, thus far, been oriented toward technology development. High capital costs and dependence on government subsidies have deterred investors. New and innovative business models have now emerged where these technologies can capture new niche markets by focusing on demand with supportive policy regimes. The papers in Part IV of this book (Chaps. 13–15) establish the feasibility of RE production and use when evaluated more comprehensively, going beyond a direct cost-benefit analysis and including environmental and climate-related externalities of traditional fossil fuel-based energy use. These papers reveal that RE is a viable option on its own, based on purely commercial factors.

## 12.1 Introduction

The world's energy supply has always been and continues to be, largely based on fossil fuels. Despite the years of attempts to reduce the dependence on fossil fuels, it is estimated that by 2030, 80 % of the primary energy mix will be dominated by fossil fuels. Oil will remain the dominant fuel, and demand for coal will rise more than that for any other fuel in absolute terms. In such a scenario, the realization that these sources of energy will not last forever and are also contributing to

P. Agarwal (✉)
Panitek AG, Vaduz, Liechtenstein
e-mail: pankaj@panitek.com

K. Mitra
InsPIRE Network for Environment, New Delhi, India
e-mail: kinsuk@inspirenetwork.org



S. Hostettler et al. (eds.), *Sustainable Access to Energy in the Global South*,
DOI 10.1007/978-3-319-20209-9_12

environmental problems strengthens the focus on renewables as a lucrative and sustainable option. This is evident, as we see governments around the globe, along with industries, thinking seriously about alternative sources of energy, a need for which was further affirmed by the 1973 oil embargo and oil price shock of 2008, coupled with fluctuating oil prices. In addition, the Fukushima disaster of 2011 has raised a big question mark regarding nuclear power.

Recent studies (IEA 2008) underscore that current global trends in energy supply and consumption are patently unsustainable—environmentally, economically, and socially. It also goes on to add that the situation can be changed if the supply of reliable and affordable energy is secured and a rapid transformation is made to a low-carbon, efficient, and environmentally benign energy supply system.

## 12.2 Renewable Energy: A Viable Alternative to Fossil Fuels

In this context, renewable energy (RE) has a significant potential to mitigate global climate change, address regional and local environmental concerns, reduce poverty, and increase energy security. Now, more than ever, countries all over the world fully recognize the need to promote the widespread adoption of RE technologies. It is also presumed that with increasing scope, scale, research, and development, the cost of RE technologies will be reduced, making it affordable and allowing REs to make major contributions to electricity generation, heating, cooling, and transport. Estimates highlight that RE could contribute at least one-half of all the electric power in each of the large economies by 2050, even in countries where electricity demand is significantly high.

However, the general approach toward RE mainstreaming has, thus far, been oriented toward technology development. While this is essential in the initial stages, the high capital costs of these technologies have deterred investors. New and innovative business models have now emerged where these technologies can capture new niche markets by focusing on demand with supportive policy regimes. The three contributions of Part IV (Chaps. 13–15) provide justifications for the application of RE technologies under different demand scenarios, and they provide clarity in the selection of appropriate technology mix to match the consumption pattern. This is essential for the development of suitable models for public–private partnerships that could lead to a proliferation of RE-based products/services.

## 12.3 New Business Models for the Mainstreaming of RE

These papers establish the feasibility of RE production and use when evaluated more comprehensively, going beyond a direct cost-benefit analysis and including environmental and climate-related externalities of traditional fossil fuel-based energy use.

We learn from a learning curve methodology-based study in South Africa that the levelized cost of electricity from photovoltaic systems, with certain assumptions based on recent trends, is expected to fall below the equivalent cost from new coal-based power stations by the end of 2016 (Chap. 13). This would indeed make photovoltaics a dominant energy source that would require an elaborate redesign of the energy generation mix and the associated systems. However, thus far, it is seen that the authorities are slow to bring about changes that are required to rapidly promote widespread rooftop solar systems. One of the reasons for this could be the loss of revenue to the government because rooftop solar provides "energy independence," which could lead to a constrained budget for providing services to the poor. The other reason is the impact that it would have on the coal industry, including the loss of jobs. Both of these are understandably genuine concerns and need to be addressed.

This piece of research has attempted to predict the impact of a large-scale switch to photovoltaics on employment, as well as revenue loss to the government. The predictions show that in the medium term, the municipal authorities would not be affected to any significant degree by revenue loss, based on a study of their budgets over the last five years. Even more surprisingly, the research finds that the net impact on employment is actually positive. Based on the above, the main inference is that the authorities should rapidly make the revisions necessary to move toward solar rooftop photovoltaic systems for energy. This would not only lead to higher employment, but would also have significant benefits in terms of reduced carbon emissions.

Based on a long and diligent initiative in India on the micro, small, and medium enterprises (MSMEs) sector, an interdisciplinary team of researchers has identified serious gaps in the current policy scenario (Chap. 14). These industries have proliferated across India over time, and they are responsible for a large fraction of production and employment in the Indian economy. However, they continue to use energy-inefficient technologies, and there are many inherent barriers, which makes it difficult for them to invest and operate with better and modern technology. Under this scenario, they are vulnerable to rising energy costs and cycles in the economy that are unfavorable to them. In addition to higher cost of energy, they have to cope with emerging pollution norms, which is obviously a challenge for those stuck with older methods of production.

However, the factors identified above provide these industries with an opportunity to use biomass-based clean energy technologies to replace high cost and polluting fossil fuels. This research uses the case of biomass thermal gasifiers in the MSME sector to explore the potential of their diffusion and scaling-up within the industry. The findings suggest that a favorable situation would require a framework for a comprehensive policy approach focused on the MSME sector, along with working with industry associations and other stakeholders in mobilizing and skill development. A cluster service approach is also found to be an effective model for faster dissemination.

Moving on to a more generalized perspective on the subject, we learn that mini-grids have the potential to provide the bulk of the solution to achieve the target of universal electricity access by 2030 (Chap. 15). In fact, the International Energy Agency (IEA 2011) concludes that in such a situation, 40 % of the energy would be produced from mini-grids. This obviously suggests that a large proportion of these mini-grids should be based on RE rather than fossil fuels. At present, it is seen that the expansion of mini-grids is almost entirely based on diesel power. The experiments with green energy-based mini-grids have remained mostly at the pilot stages.

This paper helps us understand the barriers, such as a lack of access to affordable longer term finance, weak effective consumer demand, local capacity issues, and policy uncertainty, that are preventing the development of environmentally friendly, renewables-based mini-grids. Based on the experiences in India, where village electrification is a high priority, the paper analyzes the potential for adoption and scale-up. It considers four pillars of sustainability: (1) technical, (2) financial, (3) institutional, and (4) social, as well as environmental concerns such as climate change and local impacts. To overcome the challenges in costs, management, and operations, it suggests a cluster-based approach to maintain a certain threshold in scale. This would also help to address social and institutional dimensions that are especially important when a new approach is being introduced. In terms of choice of technology and suitable technology mix among the several that are available, the researcher presents a decision-making tool for ranking different RE-based mini-grid options. The tool is based on a scoring matrix that considers capital cost, generation cost, environmental impact, local manageability and ease of operation, and the potential for enhancing livelihood opportunities as explanatory variables. This tool would be useful for both policy makers and developers.

## 12.4 Conclusions

The papers reveal that RE is a viable option on its own, based on purely commercial factors. This is in contrast to the environmental concerns regarding the rapidly increasing use of fossil fuels that initially brought the need for RE to the fore. Therefore, combining the two benefits, the commercial gains from switching to renewables and the significant impact it would have on reducing greenhouse gas emissions, it is indeed a win-win option. However, we are still in the process of understanding the policy thrusts that are needed to change from an old and established fossil fuel-based production system, as well as the choices of technology, delivery systems, cost recovery arrangements, and several other links that form part of the chain. These papers, we hope, would add to the knowledge base developed thus far and help to develop robust models that will see sustained success on the ground.

## References

International Energy Agency (IEA). (2008). *World energy outlook*. Paris: OECD/IEA.
International Energy Agency (IEA). (2011). *World energy outlook*. Paris: OECD/IEA.

# Chapter 13
# Local Government Resists the Implementation of Renewable Technologies

**David R. Walwyn**

**Abstract** South African local government, anxious to protect their revenue streams, contends that photovoltaic (PV) energy will be a disruptive and retrogressive technology, threatening the delivery of services to poor communities and causing job losses within the energy sector. In this paper, a quantitative approach is used to proportion the extent of the revenue loss, its impact on service provision, and the net outcome on employment. Using financial information and profiles of electricity consumption from two major metropolitan areas and industry-wide data on PV jobs, it is concluded that contrary to the prevailing sentiment, PV technology could stimulate economic growth, create employment, and reduce inequality. Given that the levelized cost of electricity (LCOE) from PV is predicted to fall below the equivalent cost from new coal-based power stations by the end of 2016, local authorities will be well advised to drop their resistance toward rooftop solar and take proactive steps to reap the dividends of the green economy, while making the necessary provisions to safeguard their redistributive programs.

## 13.1 Introduction

South Africa's Renewable Energy Independent Power Producers Procurement Programme (REI4P) has transformed its energy sector (Eberhard 2013; Walwyn and Brent 2015). Prior to the launch of the REI4P in 2012, electricity generation was performed by a single-state-owned entity, Eskom, which was also responsible for bulk electricity distribution, mostly to local authorities but also directly to rural consumers and large industry (Tsikata and Sebitosi 2010). The municipalities in turn have retailed electricity to a range of suburban consumers, often at a significant markup to the wholesale supply price from Eskom (Trollip et al. 2012). This margin

D.R. Walwyn (✉)
Department of Engineering and Technology Management, University of Pretoria, Pretoria, South Africa
e-mail: david.walwyn@up.ac.za



S. Hostettler et al. (eds.), *Sustainable Access to Energy in the Global South*,
DOI 10.1007/978-3-319-20209-9_13

has become an important revenue source, and it is claimed that these funds are used to provide electricity at lower cost to the poor and to subsidize other services such as water, road maintenance, and refuse removal.

Subsequent to the implementation of the REI4P, electricity generation has become more diversified. A total of 64 independent power producers, collectively responsible for 3.9 GW of new power, have been licensed by the National Energy Regulator of South Africa (NERSA) and are in the process of completing, or have already completed construction of their generation facilities. Although the program has not directly affected electricity distribution, its facilitation of the renewable energy sector has significantly influenced the economics of small-scale power generation and in particular rooftop solar. Rising costs of energy from the national grid, driven by the construction of two new coal-based power stations (Medupi and Kusile), and falling costs of power from photovoltaics (PV), indicate that rooftop solar could reach grid parity by the end of 2016. Furthermore, local generation offers a broad range of additional benefits including independence from an increasingly unreliable national grid, local employment creation, and the obvious favorable impact on carbon release and mitigation of climate change.

Nevertheless, local authorities have been recalcitrant in the development of the appropriate systems and regulations supporting the implementation of widespread rooftop solar. The justification for this approach is cited as its negative impact on jobs and municipal income. In terms of the latter, it is contended that a shift to off-grid market generation will change the extent of grid-based distribution through local authorities and hence undermine a major revenue stream. For instance, Trollip et al. (2012) predict "a decrease of between 17 and 25 % in electricity revenue collection from the residential sector alone and significant additional potential losses driven by similar dynamics in the commercial sector." Furthermore, it is posited that the erosion of coal-based energy supply will impact on the overall job market in South Africa, with employment being shifted from local coal mines to countries manufacturing low-cost solar panels.

In this paper, the extent to which rooftop solar will affect municipal income and local employment is proportioned and discussed, following which a number of recommendations are made as to how local and national authorities could adjust to the likely changes in energy generation. The paper highlights an apparent policy conflict within South Africa and presents a possible resolution to ensure that both pro-poor and pro-environment initiatives are sustained.

## 13.2 Overview of Photovoltaic Energy Generation

### *13.2.1 Cost of Photovoltaic Energy*

Unlike the cost profiles of most fossil fuels and particularly coal, the cost of electricity generation from renewables has fallen dramatically over the last decade and is predicted to continue this steep decline until at least 2030 (Feldman et al. 2012). The levelized cost of electricity (LCOE) from PV panels, geothermal sources, and onshore wind generators is now below grid parity for many countries and certainly below the equivalent greenfield cost of electricity from either conventional coal or nuclear technologies (see Fig. 13.1).

This ranking of electricity generation technologies is a new phenomenon; even as recently as 2012, large-scale comparative surveys have placed non-renewable technologies as more expensive options compared to nuclear and fossil fuels (Department of Energy and Climate Change 2012; Kost et al. 2012; IRENA 2013). The LCOE from solar PV installations has decreased dramatically, with the prices of PV modules falling by 60 % over the period January 2011 to December 2012 (IRENA 2012; De la Tour et al. 2013). This sharp decrease in prices is expected to continue and has become known in the industry as Swanson's Law, which states that the "price of solar PV modules tends to drop 20 % for every doubling of cumulative shipped volume" (Swanson 2006; see Fig. 13.2). Similarly, the LCOE cost of wind power is competitive even against gas-fired generation with some of

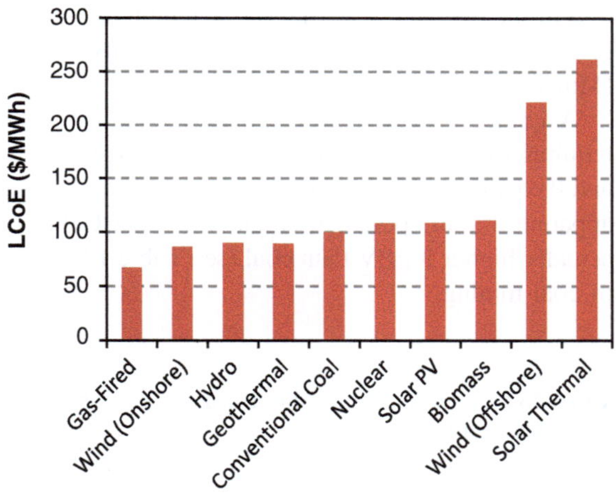


**Fig. 13.1** A comparison of LCOE by technology, updated for changes in PV module prices as of September 2013 (*source* of data U.S. Energy Information Administration 2013)

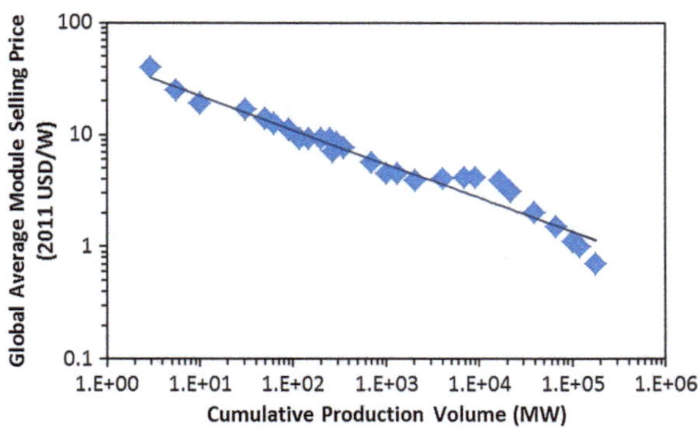


**Fig. 13.2** PV module selling prices decrease with cumulative production (data adapted from IRENA 2013)

the lowest-cost producers being in the range of US$40–50/MW (IRENA 2012). Installed overnight costs of wind turbines are about US$1800–2200/kW in most countries with the exception of China where costs are 50 % cheaper (IRENA 2012).

### *13.2.2 Employment for Photovoltaic Energy*

Renewable energy technologies (RETs) have significant employment potential (Sawin 2003; REN21 2013), as shown in Table 13.1, although these jobs have become highly localized as some countries have been more successful than others in growing sustainable sectors.

In terms of PV, it is reported that on average about 30 jobs are created per peak MW installed capacity, with the jobs being located mostly in the assembly of PV modules and their subsequent installation (Ban-Weiss et al. 2004). Excluding installation, distribution, and integration, more than 54 % of the manufacturing jobs are located in module assembly (Mulcahy 2012), equivalent to 11 jobs per peak MW. On the other hand, labor costs for this portion of the value chain account for 6 % of the overall costs, suggesting a high ratio of employment creation to cost and consequently a large potential opportunity for low-cost employment creation (Mulcahy 2012).

A comparison of jobs created per gigawatt hour (GWh) of energy generated requires some assumptions about capacity factors, coal quality, level of localization, and the extent of backward integration. Using average figures for South Africa, it is apparent that the potential employment per GWh, including the mining of coal, is actually higher for PV than coal (see Table 13.2), despite the labor-intensive nature of coal mining.

**Table 13.1** Total number of direct and indirect jobs (expressed in thousands of jobs) in the renewable industry

| Technology | Global | China | Germany | India |
|---|---|---|---|---|
| Biomass | 753 | 266 | 57 | 58 |
| Biofuels | 1379 | 24 | 23 | 35 |
| Biogas | 266 | 90 | 50 | 85 |
| Geothermal | 180 | | 14 | |
| Hydropower (small) | 109 | | 7 | 12 |
| Solar PV | 1360 | 300 | 88 | 112 |
| Solar thermal (CSP) | 53 | | 2 | |
| Solar heating/cooling | 892 | 800 | 11 | 41 |
| Wind power | 753 | 267 | 118 | 48 |
| Total | 5745 | 1747 | 370 | 391 |

Source of data REN21 (2013), Wei et al. (2010)

Table 13.2 A comparison of total employment in coal versus PV electricity generation

| | Units | Coal | PV |
|---|---|---|---|
| Capacity factor (@ 8,760 h/year) | % | 80 % | 30 % |
| Employment (average) | Jobs/MWa | 1.01 | 9.0 |
| Employment (energy) | Jobs/GWh | 0.115 | 1.027 |
| Employment (raw material supply) | Jobs/GWh | 0.169 | 0.0 |
| Total jobs/GWh | | 0.284 | 1.027 |

Data for this table developed from this study and Huntington (2009)

## 13.3 Policy Trends in Local Authorities

### *13.3.1 Redistribution and Technology*

Levels of income inequality and poverty in South Africa persist despite 20 years of democracy and policy reform in favor of redistribution and inclusive development. Thirty-nine percent of the population has an annual income of less than US$1.5 per day, and the Gini coefficient is 0.69, one of the highest in the world (National Planning Commission 2011). It is therefore not surprising that greater levels of employment, increased per capita income, and growth in the share of national income by the poor remain key objectives of government policy. At the level of local authorities, the policy documents reflect similar objectives, translated into specific initiatives to broaden service delivery, expand public infrastructure, and address poverty (City of Johannesburg 2013).

Although tiered tariff structures are a feature of the broadening of service delivery, government's approach to economic development is a mixture of direct redistribution and new wealth generation through the enhancement of overall capabilities, technological innovation, and economic freedom. The country has seemingly embraced the principles of innovation-led growth and the greater use of technology in all aspects of economic activity, although labor influence within government and large increases in social grants are undermining business confidence in the economy and constraining pro-employment investment.

### *13.3.2 Green Economy*

The prospects of the potential economic benefits from a green economy (economic growth, employment, and energy efficiency) and the global pressure on South Africa to reduce carbon emissions have prompted local authorities to adopt pro-green strategies, including stimulating the manufacture and uptake of solar geysers, the assembly and installation of PV panels, development of municipal biogas, solar farms on mine dumps, recycling projects, and the implementation of energy-efficient building codes (City of Johannesburg 2013). While these initiatives have direct environmental benefit, they also have indirect economic benefits, thereby ensuring a double dividend for municipalities and national authorities.

Notwithstanding the obvious benefits thereof, local authorities have been reluctant to allow the installation of rooftop solar and its connection to the local grid, even though systems of less than 100 MW capacity do not require licensing from the NERSA. Most importantly, the necessary systems and regulations, such as the implementation of integration standards and an appropriate feed-in tariff, have not been developed, neither is there any discussion on when these systems and regulations will be in place.

It is widely perceived that this shift to PV and a low-carbon economy presents local authorities with a significant challenge, namely balancing their own revenue requirements with broader policy goals of economic development. In particular, the widespread adoption of rooftop solar and other means of local power generation could undermine revenue collection and hamper other programs on infrastructure development and redistribution. In the remainder of this paper, the structure of tariffs, patterns of electricity consumption, and the extent to which the latter could be impacted by rooftop solar are discussed in more detail.

## 13.4 Electricity Tariffs

### *13.4.1 Predictions on Photovoltaic Uptake*

As noted in the introduction, local authorities purchase electricity in bulk from Eskom and then retail energy to a broad range of urban consumers using a regulated tariff structure. The present structure allows for some subsidization of electricity purchases by the low-energy users (and hence presumably the poor), but the extent to which this is taking place has not been widely analyzed or published.

Based on data from Cape Town Municipality, it is apparent that domestic consumers in the category 0–500 kWh/month, who collectively constitute about 66 % of the total consumer and account for 37 % of the total energy consumption, receive a net subsidy of R27 million,[1] equivalent to the small sum of R94 per household per year (see Figs. 13.3 and 13.4). On the other hand, consumers in the highest usage category (>3000 kWh/month) represent 0.2 % of total consumers and account for 2 % of total consumption, but pay a net subsidy of about R9000 per household per year.

The most significant barrier to the adoption of green energy technologies is cost (Zhang et al. 2011). However, perceptions of the initial investment are mediated by monthly energy expenditure, with higher-usage consumers being more likely to

[1]The exchange rate is about 12 South African rand (R) = 1 United States dollar (US$).

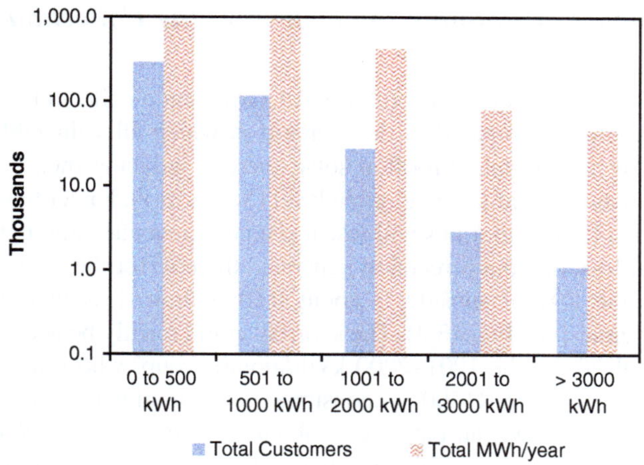


**Fig. 13.3** Profile by usage of City of Cape Town electricity consumers (number of consumers per category and total energy consumed)

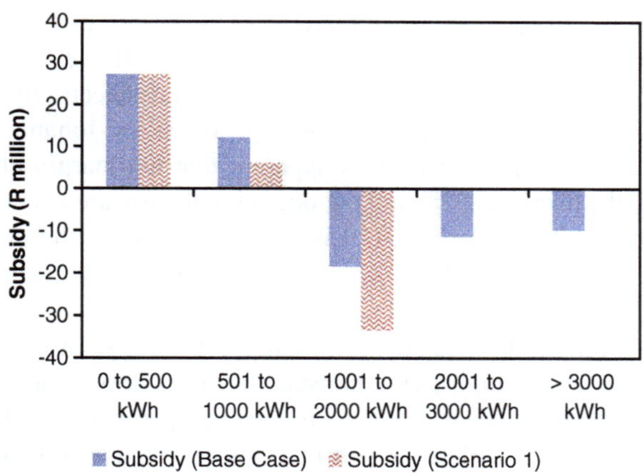


**Fig. 13.4** Subsidy levels per year for base case and Scenario 1

take a longer-term view of return on investment as a consequence of an upfront decision to purchase capital equipment such as PV panels or solar water heaters. For example, high-usage consumers may consider a return on investment of 5–10 % over a ten-year period to be acceptable, as opposed to lower-usage consumers who will not invest unless the project achieves a maximum payback period of two to three years. On this basis, it is predicted that consumers in the category of greater than 2000 kWh/month (referred to as "vulnerable consumers") are more likely to switch to rooftop solar. In the next section, the implication of this change on the customer base of local authorities is explored.

### *13.4.2 Impact of Rooftop Solar on Revenue Collection*

In the analysis of the impact of rooftop solar on revenue collection, the base case (present situation) is compared to the scenario in which all vulnerable consumers are assumed to have installed rooftop solar, thereby reducing their energy dependence on the national grid to a negligible level (Scenario 1). Under this assumption, it is evident that local authorities will lose a source of revenue, although the amount is a great deal less than the perception generated thereof; for the case of the City of Cape Town, the loss of subsidy is about R20 million (less than 1 % of total electricity revenues; see Fig. 13.4). The shortfall could readily be accommodated by shifting the subsidy to the 1001–2000 kWh/month group, whose level on subsidy will double from R660 to R1200 per consumer per year, although the impact on the actual tariff will be small (an increase of about 3 % from 92.5 to 96 cents/kWh).

This analysis shows that a shift by high-end consumers to rooftop solar can be accommodated by a small adjustment to electricity tariffs in order to retain pro-poor service delivery. It has also been argued, however, that rooftop solar could also destabilize the overall finances of local authorities, which is ultimately a question of profitability or earnings from electricity sales rather than revenue. Generally, municipalities do not separately track the cost of electricity distribution, the figure being amalgamated with the overall cost of services. However, in the case of Johannesburg, electricity is provided through a separate electricity utility (Joburg Power). Based on a comparative analysis of Cape Town and Johannesburg financial statements over the last five years, it is apparent that net margin after operational costs is 20 % of the average tariff, and the operational costs are about 13 %. In other words, if the total average tariff is R1/kWh, 67 cents covers the purchase of bulk electricity and 20 cents is used for the costs of distribution, leaving a net profit of 13 cents.

It is now possible under Scenario 1 to calculate the loss in earnings as a consequence of the shift by high-usage consumers to rooftop solar. Although these customers are paying between 15 and 20 % more per kWh, their relative consumption is low (<5 %), with the result that the net impact on earnings will be a decrease of 5 % (R11 million in the example of Cape Town). Within a scenario of increasing demand and hence high electricity sales/revenue, this loss of earnings could be readily accommodated within the city budget.

## 13.5 Policy Recommendations

Local authorities in South Africa have adopted a range of policy measures to address what are widely considered to the most critical social needs (employment and income equality), including mechanisms to facilitate redistribution and stimulate the green economy. However, these same authorities have failed to respond to the rising potential of rooftop solar on the basis that widespread importation of PV systems would destroy local jobs and that revenue losses from high-end consumers would jeopardize pro-poor initiatives.

In this paper, it is shown that both contentions are unfounded. An analysis of both income and employment data has revealed that the loss of earnings or revenue as a result of the shift will be minimal and will probably be absorbed by an overall increase in energy consumption as a consequence of the growing customer base. Moreover, the experience of the REI4P has confirmed that carefully managed procurement can result in the net addition of jobs, even at relatively low levels of local content (about 50 %), and at about the same average levels of similar programs in other countries. In other words, the stimulation of rooftop solar within South Africa's cities would result in the following important benefits:

- Creation of sustainable employment in the local green economy (one job per GWh);
- Greater capacity within local authorities to deliver services to poor communities as a consequence of growing energy independence within the high-usage consumers; and
- Lower-carbon footprint.

Furthermore, these benefits will accrue at relatively minimal financial impact on the local authorities, including revenue, earnings, and capacity to support tiered tariff structures.

The updated Integrated Resource Plan for Electricity (Department of Energy 2013) notes that rooftop solar could add as much as 20 GW of additional capacity to South Africa's overall energy generation by 2028 (20 % of total capacity). Clearly, national government is aware of the imminent PV revolution; it now needs to act by designing and implementing both systems and feed-in tariffs that will allow individual investors the maximum possible benefit from their rooftop PV. Furthermore, local authorities also need to respond by ensuring that this change happens to their advantage. Some of the more important steps that such authorities could take are as follows:

- Careful financial planning to avoid any negative impact of the relatively minor changes in their revenue streams on the sustainability of the city in general and more specifically on their pro-poor initiatives;
- A detailed analysis of the PV supply chain to identify segments that could benefit from public sector investment (such as research and development or training) and could be extended in order to maximize the local content of rooftop solar systems; and
- Reconsideration of their present negative toward energy independence, which is impeding the implementation of the technology and hampering, sustained investment by technology-based companies in the development of the necessary hardware and software of rooftop solar installations and networks.

## References


Ban-Weiss, G., Larsen, D., Li, S. X., & Wilusz, D. (2004). *Job creation studies in California for vote solar*. University of California, Berkeley. www.votesolar.org. Accessed March 9, 2015.

City of Johannesburg. (2013). *2013/16 Integrated development plan: Implementing the Joburg 2040 strategy*. City of Johannesburg. www.joburg.org.za/images/stories/2013/March/March2/2013-16%20IDP.pdf. Accessed March 9, 2015.

De la Tour, A., Glachant, M., & Ménière, Y. (2013). Predicting the costs of photovoltaic solar modules in 2020 using experience curve models. *Energy, 62*, 341–348.

Department of Energy. (2013). *Integrated resource plan for electricity 2010 to 2030; Updated report 2013*. Department of Energy (Pretoria). http://www.doe-irp.co.za/. Accessed March 9, 2015.

Department of Energy and Climate Change. (2012). *Electricity generation costs*. Department of Energy and Climate Change, United Kingdom. www.gov.uk/government/publications/electricity-generation-costs. Accessed September 14, 2013.

Eberhard, A. (2013). Feed-in tariffs or auctions? Procuring energy supply in South Africa. *Viewpoint,* Note Number 338. The World Bank Group (Washington).

Feldman, D., Barbose, G., Margolis, R., Wiser, R., Darghouth, N., & Goodrich, A. (2012). *Photovoltaic (PV) pricing trends: Historical, recent, and near-term projections*. National Renewable Energy Laboratory (NREL), Golden, Colorado.

Huntington, H. G. (2009). Creating jobs with 'green' power sources. *EMF OP 64*. Reprinted from USAEE Dialogue 17(1), March 2009. https://web.stanford.edu/group/emf-research/docs/occasional_papers/OP64.pdf. Accessed March 9, 2015.

IRENA. (2012). *Summary for policy makers: Renewable power generation costs*. International Renewable Energy Agency (IRENA), Bonn. http://www.irena.org/DocumentDownloads/Publications/Renewable_Power_Generation_Costs.pdf. Accessed March 9, 2015.

IRENA. (2013). *Renewable energy innovation policy: Success criteria and strategies*. IRENA Working Paper. International Renewable Energy Agency (IRENA), Bonn. http://www.irena.org/menu/index.aspx?mnu=Subcat&PriMenuID=36&CatID=141&SubcatID=322. Accessed March 9, 2015.

Kost, C., Schlegl, T., Thomsen, J., Nold, S., & Meyer, J. (2012). *Levelized cost of electricity renewable energies*. Renewable Energy Innovation Policy Group, Fraunhofer Institute for Solar Energy Systems. http://www.ise.fraunhofer.de/en/publications/veroeffentlichungen-pdf-dateien-en/studien-und-konzeptpapiere/study-levelized-cost-of-electricity-renewable-energies.pdf. Accessed March 9, 2015.

Mulcahy, M. (2012). *Review of the competitive bid for PV in South Africa: Is SA maximizing job creation and value for money from its photovoltaic industry?* Research paper presented to the Graduate School of Business University of Cape Town.

National Planning Commission. (2011). *National Development Plan 2030: Our future—Make it work*. The Presidency, Republic of South Africa (Pretoria). https://www.dac.gov.za/sites/default/files/NDP%202030%20-%20Our%20future%20-%20make%20it%20work_0.pdf. Accessed March 9, 2015.

REN21. (2013). *Renewables 2013: Global status report*. Renewable policy network for the 21st Century (REN21). http://www.ren21.net/REN21Activities/GlobalStatusReport.aspx. Accessed March 9, 2015.

Sawin, J. (2003). *Policy lessons for the advancement & diffusion of renewable energy technologies around the world*. Thematic Background Paper, The Energy and Resources Institute, Worldwatch Institute (Bonn). http://www.worldfuturecouncil.org/fileadmin/user_upload/Miguel/Sawin__2004__National_policy_instruments.pdf. Accessed March 9, 2015.

Swanson, R. M. (2006). A vision for crystalline silicon photovoltaics. *Progress in Photovoltaics: Research and Applications, 14*, 443–453. doi:10.1002/pip.709.

Trollip, H., Walsh, V., Mahomed, S., & Jones, B. (2012). Potential impact on municipal revenue of small scale own generation and energy efficiency. In *Conference Paper Submitted to the South African Economic Regulators Conference (Pretoria).* www.nersa.org.za/SAERC/Documents/SAERC/Hiltop%20Trollip.pdf. Accessed March 9, 2015.

Tsikata, M., & Sebitosi, A. (2010). Struggling to wean a society away from a century old legacy of coal based power: Challenges and possibilities for South African Electric supply future. *Energy, 35*(3), 1281–1288.

U.S. Energy Information Administration [eia]. (2013). *Levelized cost of new generation resources in the annual energy outlook 2013*. U.S. Energy Information Administration (Washington).

Walwyn, D. R., & Brent, A. C. (2015). Renewable energy gathers steam in South Africa. *Renewable and Sustainable Energy Reviews, 41*(1), 390–401. doi:10.1016/j.rser.2014.08.049.

Wei, M., Patadia, S., & Kammena, D. M. (2010). Putting renewables and energy efficiency to work: How many jobs can the clean energy industry generate in the US? *Energy Policy, 38*, 919–931.

Zhang, X., Platten, A., & Shen, L. (2011). Green property development practice in China: Costs and barriers. *Building and Environment, 46*(11), 2153–2160.

# Chapter 14
# Green Mini-grids: Evidence from India's Experience Provides Lessons for Scale-up in Low-Income Countries

**Ritu Bharadwaj and Somnath Bhattacharjee**

**Abstract** The International Energy Agency (IEA) has estimated that, in order to achieve universal access to electricity, mini-grids will have to provide around 40 % of the new capacity needed by 2030. While the use of diesel mini-grid systems is expanding, green mini-grid options exist, but have struggled to expand beyond pilot projects and will not make a meaningful contribution to sustainable energy for all targets at the current deployment rates. Key barriers include lack of access to affordable longer-term finance, weak effective consumer demand, local capacity issues, and policy uncertainty. This paper presents the experience and lessons of policy initiatives in India to promote village electrification based on renewable-energy-powered mini-grids that have high scale-up potential worldwide, especially in Africa. An analysis of the scale-up potential of the renewable-energy-based mini-grids implemented in India has been undertaken under the four pillars of sustainability, viz. (1) technical, (2) financial, (3) institutional, and (4) social. In addition, the assessment also covers environmental sustainability to ensure that the chosen options are not only climate–friendly, but also do not lead to local environmental problems. This assessment has been used to suggest a strategy for scale-up in low-income countries particularly in Africa, which includes a cluster-based approach to mini-grids that can help in providing the scale of operation to cover the issues around the management, operation, cost recovery, social and institutional issues, and diversification of energy use. A decision-making tool for ranking various renewable-energy-based mini-grid options has also been presented, which includes a scoring matrix based on the capital cost, generation cost, environmental impact, local manageability, and ease of operation and the potential for enhancing the livelihood opportunities.

R. Bharadwaj (✉) · S. Bhattacharjee
Institute for Industrial Productivity, New Delhi, India
e-mail: ritu.bharadwaj@iipnetwork.org

S. Bhattacharjee
e-mail: somnath.bhattacharjee@iipnetwork.org



S. Hostettler et al. (eds.), *Sustainable Access to Energy in the Global South*,
DOI 10.1007/978-3-319-20209-9_14

## 14.1 Introduction and Purpose

More than 20 % of the global population lacks access to electricity (BP 2012), and almost twice the same are deprived of clean sources of cooking. The majority of these people live in remote rural locations with a poor quality of life, low per capita income, and limited access to basic services such as healthcare, sanitation, nutrition, and so forth, which is quite often constrained due to lack of access to electricity. This huge gap in clean energy supply and remoteness of locations imply that electricity will have to be provided through both centralized systems, that is, grid-based supply and decentralized energy systems including mini-grids and off-grid solutions. Here, mini-grids offer a better and more sustainable solution—they can be deployed more rapidly and with less investment than grid systems that require a much larger and longer-term effort; they can be easily taken to remote, rural locations; and they can be customized to suit the local requirements and livelihood generation options. The International Energy Agency (IEA) has estimated that, "to achieve universal access to electricity, 70 % of the rural areas that currently lack access will need to be connected using mini-grid or off-grid solutions." (IEA 2012b).

India is faced with a similar situation, where serious efforts have been made by the government to address the issue of rural electrification through programs such as Remote Village Electrification[1] (RVE), Village Energy Security Program[2] (VESP), and the Rajiv Gandhi Grameen Vidyutikaran Yojana (RGGVY). These programs promote both grid and mini-grid options through a range of stakeholders that include nongovernmental organizations (NGOs), local government units, community groups, and franchisees of distribution utility. But, despite being successful in certain aspects and in selected geographical regions, the challenges still remain formidable. Nonetheless, these initiatives provide several lessons that can be useful for scaling-up such efforts not just in India but in other parts of the world.

[1]Government of India, Ministry of New and Renewable Energy (MNRE) is implementing the RVE program that provides financial support for electrification of remote unelectrified villages and unelectrified hamlets of electrified villages. Such villages are provided basic facilities for electricity/lighting through various renewable energy sources. Small hydropower generation systems, biomass gasification-based electricity generation systems, solar photovoltaic power plants, etc., in distributed power generation mode/mini-grids are used depending upon the availability of resources for generation of electricity.

[2]The VESP was launched by MNRE, with a view to explore the development of community-driven biomass-based resources to meet the basic energy needs of the remote villages, where electrification is not possible.

## 14.2 Current Status of Energy Access and Mini-grids in India

Since independence, the Indian power sector has grown manifold in size and capacity. The installed generation capacity, which was 1364 megawatt (MW) in 1947, now stands close to 255,012 MW (CEA 2014). The village electrification during the same time grew from a mere 1500 villages to more than 560,000. The structure, ownership patterns, and regulatory setup of the power sector have witnessed radical changes, especially in the past few years as part of the ongoing reform program, with the establishment of the independent regulators, corporatization, unbundling, and entry of private players in some states. However, in spite of this impressive achievement, the power industry continues to be under constant pressure to bridge the gap between the supply and demand of electricity. "In 2005, a total of 412 million people in India had no access to electricity, with 380 million of them (92 % of total population) living in rural areas and 32 million in urban areas. According to recent IEA estimates, India is today 64.5 % electrified, with an urban electrification rate reaching 93.1 % and a rural rate of only 52.5" (IEA2012a, b). Others with access often have to cope with poor and erratic availability of electricity and other fuels. With constraints faced in resource availability and in delivery mechanisms, traditional means of energy supply are falling short. Improving access to electricity for far-flung villages poses additional problems. There are many villages that are so remotely located so that grid connectivity is either not feasible or not cost effective.

To tackle this challenge in an environmentally friendly manner, the Government of India has given a strong thrust to renewable-energy-technology-based mini-grids wherein power can be generated with the help of local energy resources and those resource-based proven technologies. Numerous renewable and small-scale mini-grid technologies have been developed to the stage where their technical feasibility has been proven and they have been adopted in their full scale worldwide. The Electricity Act (2003) has also given a thrust to distributed generation particularly in the context of rural electrification. The Act, in addition to grid extension as a mode for rural electrification, specifies mini-grid-based generation and supply through stand-alone conventional and renewable energy systems.

But even with this strong policy push, to date, mini-grid options have struggled to expand beyond pilot projects and are not expected to make a meaningful contribution at current deployment rates.

## 14.3 Mini-grid Sustainability Issues or the Lack of It—Lessons from Earlier Programs

The mini-grid projects have many unique characteristics, which makes them distinct from normal rural electrification projects based on grid extension. Though many of these features weigh in the favor of mini-grids, if we review it based on its cost, ease of installation, and so on, there are a host of issues at the local level normally not considered in the project development stage that are responsible for the failure of many of these projects. The successful implementation and penetration of mini-grids largely depend on how effectively these issues are dealt with at the local level.

In this section (Sect. 14.3), a review of the various issues faced by mini-grids has been made under the four pillars of sustainability, viz. (1) technical, (2) financial, (3) institutional, and (4) social. In addition, a review of environmental sustainability has also been carried out to ensure that the chosen options are not only climate-friendly but also helpful in dealing with local environmental problems. The review is based on lessons from earlier government programs as well as the experience of local NGOs in dealing with renewable energy projects.

### 14.3.1 Technical Sustainability

The technical sustainability of a typical renewable-energy-based village mini-grid unit is defined by the issues that need to be addressed to reduce the plant downtime. These are:

- reliability of the chosen technology;
- adequate availability of energy resources;
- local manageability of the technology in terms of its day-to-day operation as well as periodic maintenance requirements;
- local availability of spare parts/servicing needs; and
- disposal/management of wastes generated.

Field experiences from India clearly indicate that addressing the technical sustainability issue is most essential for sustainability of any mini-grid project to operate and provide uninterrupted supply. Once this is established, it can help in building confidence and credibility with the villagers about the technology and the project developers, which can directly influence their willingness to pay for the services and recovery of the cost by the developer. However, so far, only a handful of implemented mini-grid projects in India have attained an acceptable level of technical sustainability regarding the issues listed above, and that has been a major cause for concern, which the green mini-grids program must target to address.

### 14.3.2 Financial Sustainability

The important factors/issues that govern the financial sustainability of mini-grids are as follows:

- recovery of user charges/payment for energy services;
- running cost of the chosen technology, including the energy and maintenance cost;

- value-added by-products, if any;
- load at village level is small, seasonal, mostly single shift, and fluctuating. They are thus not able to provide enough annual generation of electricity which can make the power plants profitable; and
- promotion of income generation activities, either by augmenting farm income or by setting up small, need-based rural enterprises.

Unfortunately, none of the mini-grids implemented so far in India have been able to run on a financially sustainable basis. These projects have failed to recover even the operating cost of the plants let alone the cost of recovery of capital investment. Although there are few isolated cases where recovery of cost has been impressive, these are run with NGOs' support to carry out extensive social mobilization campaigns to address the issue of financial sustainability.

### 14.3.3 Institutional Sustainability

There are too many stakeholders in the ecosystem of mini-grids. To the extent possible, there is a need to avoid involvement of multiple agencies, as also to clarify the roles and responsibilities of each involved. The success of any mini-grid project depends mainly on the commitment of the implementing agency and the resources that they can spare. Many of the existing mini-grid units are being managed by government agencies such as the state nodal agencies of Ministry of New and Renewable Energy (MNRE) and the forest department. Most of these mini-grids have failed, because they are not able to carry out the intensive social engagement process that is necessary for these projects with their limited manpower. The best role these agencies can play is in facilitation of the entire process by helping in administrative compliance/allotment of land and so forth for private sector players willing to invest.

The role of the technology suppliers is vital, and they should be made accountable not only for the maintenance of the plants, but also in the case of projects based on biomass, in the management of the supply chain of raw material. An appropriate financial model (in a Built Own Operate mode) needs to be worked out comprising of a system of penalties and rewards. In summary, the green mini-grid program must draw on the existing experience from India, so that the chance of sustaining the initiatives under the program improves.

### 14.3.4 Social Sustainability

Mini-grid systems have to be treated differently, and there has to be a strong component of community mobilization. Even if the systems are being set up on a Built Own Operate basis and the day-to-day technical and operational problems are addressed, the engagement and involvement of the village community is of critical importance. This would not only help in ensuring recovery of payment for the electricity services, but also in the integration of the needs/aspirations of the villagers in the plan for electrification, so that there is local ownership of the facilities being set up as part of the green mini-grids.

Provisions have to be kept in the green mini-grid program for social engineering, and sufficient time has to be allowed until the villagers come to terms with the new system. The program could also explore the possibilities of leveraging financial support (including exploring carbon market possibilities) to fund/support the social engineering component.

## 14.4 Strategy to Ensure Sustainability

The failure of mini-grid systems does not mean that the model is unviable. They may have been rendered dysfunctional due to one or several of the sustainability issues highlighted in the previous section. Many of these mini-grid projects can be revived by addressing these issues or with small improvements/modifications in the existing model, as detailed below.

### *14.4.1 Risk Assessment Framework and Decision Tool*

The government and private sector investors involved in a mini-grid program need support to undertake an effective risk assessment through a decision-making tool that helps them in technology selection and assessing the scale of investment. For this, it is necessary to address issues such as capital cost, operation and maintenance cost, environmental impact, local manageability, and the potential of generating livelihood opportunity to promote productive work.

A mini-grid project could be based on either conventional or renewable forms of energy. The choice of technology would depend on the appropriateness of the chosen technology for specific villages. A list of such options for green mini-grid is presented below:

- Diesel generating (DG) sets powered by biofuels (nonedible vegetable oils such as Jatropha and Pongamia),
- DG sets powered by producer gas generated through biomass gasification (100 % producer gas engines)[3],
- Solar photovoltaic (SPV), and
- Small hydro.

[3]To be considered only where project design includes dedicated energy plantations to ensure sustainable biomass supply.

It may be noted that the above list is based on the technologies that are presently being employed and are the preferred options for mini-grid-based power generation. There could be additional possibilities, which are not popular now, but may become relevant in the future, such as DG sets powered by biogas (from animal waste), wind hybrid systems, and other hybrid options, including any new technology.

Although diesel is the most convenient form of decentralized power generation option, it would be advisable to treat the diesel option as only for standby or to cope with temporary disruption in the supply of local renewable energy sources.

Figure 14.1 provides a technology decision tool that can be used as a guide for the choice of technology in particular villages, which can be further vetted through a detailed survey of the village. The tool was developed by InsPIRE Network for Environment (formerly known as Winrock International India) as part of their effort to help the Government of India come out with guidelines for renewable-energy-based mini-grids under the RGGVY program and other livelihood opportunities that it offers. This can help the decision makers arrive at a particular technology based on their needs and local conditions. Table 14.1 ranks the preferred technology options. The ratings shown in the table are indicative, but

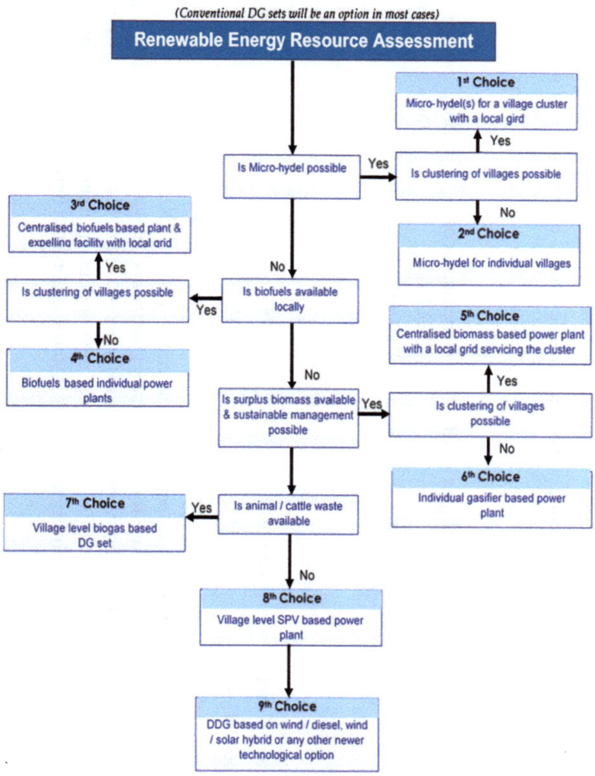


**Fig. 14.1** Technology decision tool

**Table 14.1** Framework for ranking various renewable-energy-based mini-grid options

| | Score (in a scale of 1–5, please refer to Table 14.2 for the scoring criteria) | | | | | | | Overall score | Ranking |
|---|---|---|---|---|---|---|---|---|---|
| | (1) Capital cost (INR[a]/kW) | | (2) Generation cost (INR/kWh) | | (3) Env. impact | (4) Local manageability | (5) Livelihood opportunity | | |
| | INR/kW | Score | INR/kWh | Score | | | | | |
| Biomass gasifier/DG sets | ~78,000 | 3 | 2.25 | 4 | 2 | 2 | 4 | 15 | 3 |
| SPV | ~300,000 | 1 | 14.5 | 1 | 4 | 2 | 1 | 9 | 5 |
| Biogas DG sets | ~85,000 | 3 | 0.75 | 5 | 3 | 1 | 3 | 15 | 4 |
| Biofuel DG sets | ~20,000 | 5 | 10.75 | 2 | 3 | 3 | 5 | 18 | 2 |
| Micro-hydel | ~60,000 | 4 | 0.25 | 5 | 5 | 3 | 3 | 20 | 1 |

[a]Conversion factor for Indian Rupee (INR) currency—US$1 = INR 60

*Note* Although the overall score for the biomass gasifier and biogas options is similar, gasifier-based systems are given a higher ranking, as they are fairly established as a mini-grid technology, and currently are a preferred option compared to biogas-based engines. The above scoring matrix excludes options such as wind/solar, wind/diesel hybrids, or any other new options, as mini-grid systems based on such technologies are presently not operational and there is no basis for comparing the effectiveness of such systems against the options listed above

**Table 14.2** Scoring criteria used for ranking renewable-energy-based mini-grid options in Table 14.1

| | | Scoring |
|---|---|---|
| 1 | Capital cost is the cost of power-generating unit | <INR 25,000—5 |
| | | INR 25,000–INR 75,000—4 |
| | | INR 75,000–INR 100,000—3 |
| | | INR 100,000–INR 150,000—2 |
| | | INR 200,000—1 |
| 2 | Generation cost is the operation and maintenance cost of unit power generation. For SPV systems, it refers to the cost of replacement of battery bank every four years. For biofuels-based plants, the cost of generation would be INR 9.45/kWh if the sale of press cake is also accounted for | <INR 1.5—5 |
| | | INR 1.5–INR 5—4 |
| | | INR 5–INR 7.5—3 |
| | | INR 7.5–INR 12—2 |
| | | >INR 12—1 |
| 3 | Environmental impacts include local as well as global air pollution, afforestation/deforestation impacts, solid as well as liquid waste generation | Least polluting—5 |
| | | Most polluting—1 |
| 4 | Local manageability refers to the ability of maintaining/managing the equipment/systems in remote places and their serviceability | Most robust system—5 |
| | | Least robust system—1 |
| 5 | Enhancing livelihood opportunity refers to the ability of a particular technology to promote productive work as well as local employment generation | Maximum livelihood opportunity—5 |
| | | Least livelihood opportunity—1 |

can be used as a guide for preparation of detailed project reports through local survey that can provide complete justification for selecting an option.

## *14.4.2 Anchor Load*

As highlighted in the financial sustainability section (Sect. 14.3.2), the typical loads for a mini-grid at village level are small and seasonal. They are thus not able to provide enough annual generation of electricity, which can make the power plants financially viable. Since power plants must sell above a certain threshold to generate affordable electricity, it is essential to have customers with committed base loads. To provide rapid and sustainable energy services in the rural and remote areas, leveraging the existing infrastructure would show quicker results. It would be an enormous task to supply power in isolation by way of laying a massive infrastructure of distribution and generation capacities. Typical examples for this could be the small and medium enterprises (SME) at the village level.

Having a base or anchor load for each mini-grid can help in providing a regular source of revenue to meet the cost of operation, maintenance, and supply of the mini-grid, and it can also (in some cases) help in subsidizing the electricity supply to the nearby villages. In view of electricity supply shortages in India, more than 60 % of industries already rely on captive power plants wherein huge quantities of diesel and furnace oil are already being used by all sectors.

There are around 26 million operational SME units in India, many of which are in the rural/semi-rural areas. The SMEs in rural areas can greatly benefit from green mini-grids while subsidizing power for villages. Similar to India, SMEs are a very important part of the economic activity in most of the LICs. Along with SMEs, another growing segment that could provide symbiotic benefit in this case is the telecom towers. India has more than 250,000 mobile phone towers that consume 3–5 kW of power depending on the number of operators using the tower. Many of these towers have been built in areas away from the grid and are powered by expensive diesel fuel. Even in areas where grid connection is available, due to irregular supply, the tower companies end up running backup power for almost 80–90 % of the time. These towers consume about two billion liters (about 530 million gallons) of diesel every year (TRAI 2011). To begin with, these telecom towers could be considered as viable anchor load for the green mini-grids.

### *14.4.3 Carbon Benefits*

Renewable-energy-based mini-grids for rural electrification projects have considerable potential to avail carbon credits under the clean development mechanism (CDM) as well as voluntary markets. The carbon credits can go a long way in ensuring financial sustainability by meeting the regular operation and maintenance cost of mini-grid units. Therefore, the possibility of obtaining carbon credits for the mini-grids should also be considered in the development of action plans and wherever needed appropriate guidance and facilitation should be provided to the implementing agencies for availing such benefits.

## 14.5 Conclusion and Way Forward

Accelerated deployment of green mini-grids across Asia and Africa can help fill a critical gap in creating universal energy access, as well as displace the increasing demand for diesel and kerosene as economies expand. A substantial green mini-grids' sector would both support rural development and help countries avoid high-carbon lock in and reduce dependence on fossil fuels. Many mini-grid projects in India have successfully dealt with the barriers to large-scale deployment with the help of proper technological choices and effective risk reduction measures that make them more viable. The lessons from these projects, if effectively scaled-up in

India and other parts of the world especially in low-income countries (LICs), can help in dealing with the issue of energy access in regions where these challenges are huge. Indian models are generally entrepreneur-driven, with a lot of low-cost innovation that makes it easy for scale-up.

Increased energy access can have a transformative effect on the opportunities for people to escape from poverty. If a critical mass of green mini-grids took hold in Africa and South Asia, enabling a trajectory toward a future network of smart and clean energy nodes, there is a potential to leapfrog to a more sustainable and resilient energy system that can be substantially boosted.

## References

BP. (2012). BP statistical review of world energy June 2014. http://www.bp.com/content/dam/bp/pdf/Statistical-Review-2012/statistical_review_of_world_energy_2012.pdf. Accessed 14 March 2014.

Central Electricity Authority (CEA). (2014). Monthly review of power sector report. Government of India. http://www.cea.nic.in/reports/monthly/inst_capacity/nov14.pdf. Accessed 14 March 2014.

International Energy Agency (IEA). (2012a). *Comparative study on rural electrification policies in emerging economies*. France, Paris: International Energy Agency.

International Energy Agency (IEA). (2012b). *World energy outlook 2012*. France, Paris: International Energy Agency.

Telecom Regulatory Authority of India [TRAI]. (2011). *Consultation Paper on Green Telecommunications*. New Delhi, India: Telecom Regulatory Authority of India.

# Chapter 15
# Large-Scale Diffusion of Biomass Thermal Gasifiers in India's Micro, Small, and Medium Enterprises: Experiences and Opportunities

**Shirish Sinha, Sunil Dhingra and Daniel Ziegerer**

**Abstract** Micro, small, and medium enterprises (MSMEs) form the backbone of the Indian economy. However, MSMEs are going through a turbulent phase due to overall sluggishness in the economy and rising input energy costs. In the current energy-supply scenario, energy pricing and enforcement of pollution norms for MSMEs have created a favorable environment for rapid acceptance of biomass energy-based clean energy technologies as a potential option for replacing fossil fuels in MSMEs. This paper presents insight and analysis of the experience of promoting diffusion of biomass thermal gasifiers in MSMEs across India as a clean energy technology solution and way forward for scaling-up, based on a partnership initiated in 1994 by the Swiss Agency for Development and Cooperation (SDC) with The Energy and Resources Institute (TERI). It is recommended that there is a need for a framework for a holistic energy policy for the MSME sector to strengthen and mobilize stakeholders, such as industry associations and industries' district centers, and a model for dissemination that takes a cluster service approach.

## 15.1 Introduction

Quality energy services drive economies and societies. However, the impact of climate change can affect how much energy is produced, delivered, and consumed. In light of the climate change challenge, it becomes imperative to adopt strategies for an energy mix that leads toward a low-carbon development pathway. The strategies include development or modification of energy systems that reduce greenhouse gas emissions. In India, climate change concerns, coupled with growing

S. Sinha (✉) · D. Ziegerer
Swiss Agency for Development and Cooperation (SDC), New Delhi, India
e-mail: shirish.sinha@eda.admin.ch

S. Dhingra
The Energy and Resources Institute (TERI), New Delhi, India



S. Hostettler et al. (eds.), *Sustainable Access to Energy in the Global South*,
DOI 10.1007/978-3-319-20209-9_15

concerns about long-term energy security, depleting fossil fuel reserves and their environmental impact, have provided greater stimulus to promoting renewable energy, particularly in sectors where larger gains are possible (Sinha 2012). While grid-connected and off-grid renewable energy can contribute significantly toward meeting India's electricity deficits, the decentralized renewable-energy systems based on sustainable use of resources such as biomass have the potential to provide solutions for developing low-carbon development pathways in some of India's energy-intensive sectors such as that of the micro, small, and medium enterprises (MSMEs).

Referred to as the engines of growth, the MSMEs play a critical role in India's industrial output, exports, and employment generation. In fact, the sector provides the largest share of employment after agriculture. Around 36.2 million MSMEs, spread over 180 clusters, provide employment to approximately 80.5 million people (MoMSME 2013). Together, they account for 45 % of the total manufacturing output and 40 % of the total exports of the country. With a substantial contribution of 8 % to the country's gross domestic product, the sector also forms the backbone of the economy. Further, this sector has consistently registered a higher growth rate than the rest of the industrial sector. There are over 6000 products, ranging from traditional to futuristic, which are currently being manufactured by MSMEs in India. Owing to the sector's enormous contribution toward the growth of the economy, the ready availability of, and access to, clean forms of energy at affordable prices is particularly crucial.

Despite its enormous contribution to the growth and development of the economy, the MSME sector is going through a turbulent phase. This is primarily due to the overall sluggishness in the economy and the rising input energy costs. In recent years, the prices of all forms of energy used by MSMEs, especially fossil fuels, have witnessed a steep increase. There are many energy-intensive industries among MSMEs that still use commercial fuels such as coal, oil, and gas, or even biomass, inefficiently. In most of these industries, energy costs account for nearly 30–50 % of the total production cost. Further, these MSMEs use a wide range of locally available cheaper fuels, which, together with inefficient processes and technologies, lead to adverse environmental impacts, especially at the local level (Soni 2007). For this reason, energy efficiency/fuel switching is at the center stage for improving the competitiveness of the sector and reducing greenhouse gas emissions. Biomass is one such important source of renewable energy in India, which can generate considerable energy savings in the sector. It has been estimated that about 500 million tons of biomass, including woody biomass and agro-residues, is produced annually in the country; more than 150 million tons of this is estimated to be surplus. The surplus, if used effectively in a cleaner energy form, can contribute significantly to meeting the overall energy needs of the country in a sustainable manner.

The Government of India (GoI), through its Integrated Energy Policy, the National Action Plan on Climate Change (NAPCC), and the Twelfth Five-year Plan (2012–2017), has put considerable emphasis on the use of renewable-energy systems such as the biomass-based clean energy solutions (GoI 2008; Planning Commission 2012). These policies and plans have also identified the importance of cleaner technologies in the MSME sector for improving energy efficiency, reducing transaction cost, ensuring environmental sustainability, and making them competitive in a liberalized economy. With changing energy-sector policies and increasing energy prices, especially of fossil fuels, many MSMEs are exploring ways to switch to clean energy options (Soni 2009). The switch to clean energy options will require adherence to the stringent environmental/pollution regulations applied by the State Pollution Control Boards. Biomass-based, clean energy system based on gasification technology is one such option that has the potential to transform energy use in MSMEs in India.

Although biomass-based energy solutions have demonstrated huge potential in addressing India's growing energy concerns, they are not yet seen as a mainstream renewable-energy option in the MSME sector. This is primarily due to outdated technologies, limited biomass availability, and a lack of awareness about the existing innovations. This paper reflects on the experience of the Swiss Agency for Development and Cooperation (SDC) in partnership with The Energy and Resources Institute (TERI) in promoting the diffusion of biomass thermal gasifiers as a clean energy solution in MSMEs across India. The paper also identifies barriers and suggests some possible ways forward to accelerate/up scale the diffusion of biomass thermal gasifiers in the MSME sector.

## 15.2 SDC–TERI Biomass Energy Program: A Journey of Two Decades

In 1994, SDC initiated a partnership with TERI to support MSMEs as low-capacity end users, enabling them to improve their technologies and processes and undertake measures that lessen the impact on the environment, reduce energy costs, and improve their competitiveness in the market. One specific low-capacity end-use application was biomass energy-based technology solutions for thermal/heat applications in MSMEs (Kishore and Mande 2007). Although the basic science of gasification was well established in India, it was not until 1994 that efforts were undertaken to develop technology and integrate biomass gasifier with industrial furnaces and ovens for thermal energy generation (Fig. 15.1). In the initial years of the partnership, the focus was on developing low-cost technological solutions. This was achieved by first understanding the actual process of thermal energy needs in MSME units and pilot interventions with biomass gasifiers, and taking user feedback at the field level, and then by translating this experience into technology design.

As a result of the continuous interaction with end users, process experts, and consultants, TERI was able to develop a gasifier for a variety of thermal energy applications for different MSMEs in different clusters. Consequently, biomass gasifiers for thermal use could be successfully demonstrated in enterprises such as silk reeling, textile dying, magnesium chloride production, brick drying, spices drying, wax melting, aluminum billet heating, puffed rice making, sand drying,

**Fig. 15.1** A biomass thermal gasifier in operation in a MSME unit

cardamom curing, natural rubber processing, chemical extraction, metal smelting, and mineral processing (Kishore and Mande 2007). After the successful demonstration of technology development and implementation in MSMEs, a new phase of the partnership was initiated in 2012 to accelerate the use of biomass-based energy systems that can help small enterprises secure access to clean energy services.

The innovative approach to the dissemination of biomass gasifiers for the thermal applications in MSME clusters involves setting up a local delivery mechanism comprising of technology supplier and service provider to facilitate the commercialization in the MSME units. The dissemination approach has also addressed building human capacities and providing local employment opportunities. The highlights of the approach include identification of local gasifier manufacturers, creation of a local delivery system through an entrepreneur, capacity building of the manufacturer, and entrepreneurs and MSMEs providing processed biomass linkages along with maintenance backup facilities. As an implementing partner, TERI focused on awareness generation, demand creation, development of technology, capacity building, establishing local supply-chain delivery, and capacity building of manufacturers, local service providers, and operators in selected industrial clusters.

## 15.3 Drivers and Barriers

As a result of the SDC–TERI partnership, over 650 biomass gasifiers have so far been installed for meeting thermal energy needs of MSME clusters in the country. These replications, which are demand-driven, have established biomass gasifiers as

**Fig. 15.2** A gasifier used for baking ceramic products in Ahmedabad

a cost-effective, energy-delivery system. They have also resulted in generating positive spin-off effects within and across clusters. For example, after the success of the initial pilot intervention in one silk-reeling unit in Ramanagaram, Karnataka, there was a rapid diffusion in other silk-reeling units in the same cluster. Similar experiences have been observed with food-processing units, sand drying, and rubber-drying units. Further, the interventions have created local manufacturing and delivery services in a few MSME clusters, thus providing employment and establishing clean energy entrepreneurs. For example, in the rubber-processing units in Kerala, buoyed by the positive results of energy savings by installing a thermal gasifier in the unit, the unit owner also became a licensee and manufacturer of the gasifier and disseminated it in the rubber-processing clusters of Kerala and Karnataka. The recent energy pricing and implementation of pollution norms for MSMEs have also created a favorable environment for rapid acceptance of this technology. In Gujarat, the ceramic units are faced with the rising price of natural gas and are thus shifting to invest in thermal gasifiers to meet their thermal energy requirements (Fig. 15.2). Finally, the enforcement of stringent environmental and pollution norms has also forced the units to shift toward cleaner technologies.

The experience from the dissemination efforts shows that the current installation is just a small fraction of the potential and opportunity that exists. In other words, there exists huge untapped potential. Biomass gasifier systems can meet thermal energy capacity needs ranging from 25 kWth (kilowatt thermal) to 3 MWth (megawatt thermal) and fulfill the temperature requirements of MSME units ranging from 60 to 1000 °C. This, in turn, opens up opportunities for applications in other MSME sectors and units that have similar thermal energy processes such as silk reeling, textile dyeing, hot water/steam generation, food processing, non-ferrous metal recycling (aluminum and lead), powder coating, chemicals, foundries (allied core baking operations), glass melting, charcoal and brick making, and ceramics. In addition to these MSMEs, the gasifier can also meet energy needs of other industries where precision heating is required. However, this would require further technology upgrades, including system automation, in order to make gasifier systems more robust and user-friendly.

In addition, the experience has demonstrated the financial viability of biomass gasifiers. The payback period for an MSME unit using different types of fossil fuels is anywhere between six and nine months. This payback period, however, increases to around two years if the fuel is replaced with biomass. The experience also shows that there is a significant reduction in cost per unit of useful energy through gasification. In cases where a fossil fuel is replaced, the reduction is as much as 60–80 %, while for biomass fuels, it is 50 %. For example, for thermal applications such as water heating, hot-air generation, drying, cooking, and steam generation, the cost per unit of useful energy through gasification is 60 % cheaper than the furnace oil/diesel option and 80 % cheaper than the electricity option at the current prices of diesel, electricity, and biomass fuels. Other key economic benefits are the improved productivity of MSMEs and the subsequent improvement in the quality of end products as a result of better processes and uniform heat control (Fig. 15.3).

As mentioned above, the enforcement of stringent environmental and pollution norms in the MSME sector by regulatory agencies has emerged as a new driver to encourage the use of biomass thermal gasifiers. However, there are continuing challenges. First, the diversity of energy requirements of MSMEs means that the gasifier systems need to be *customized or tailor-made* to suit each end-use application. This impedes scaling-up. Other challenges are the *lack of appropriate local support services*—in the form of local manufacturing, supply, and maintenance—

**Fig. 15.3** A gasifier used for puffed rice production in Ahmedabad

the absence of financial linkages, and a dearth of skilled human resources. Unlike other energy carriers, organization of the *biomass supply chain* and fuel supply linkages is of critical importance. Experience shows that MSMEs are reluctant to stock biomass as it requires more space, and therefore, there is a need for local delivery mechanisms to ensure sustained availability of biomass. Another major challenge is the lack of exchange of information/knowledge or best practices between MSME units and between clusters.

## 15.4 Way Forward for Scaling-up

The paper, as a way forward, proposes a guiding framework to enable the use of biomass energy technologies in India's energy-intensive MSME sector. As the experience of SDC's partnership with TERI has shown, the biomass thermal gasifier, as a renewable-energy option, is economically viable and likely to remain so in the future. Previous and ongoing efforts have resulted in a significant improvement in the understanding of the technology, in the uptake and diffusion of the gasifier in MSMEs, and in the identification of key barriers and opportunities with respect to accelerated dissemination. However, for accelerated diffusion to take place, there are still some critical barriers that need to be addressed: Gasifier technology needs to be more user-friendly through integration of automated systems for controlling various processes and a comprehensive biomass supply-chain analysis (covering issues of renewable biomass) to ensure sustained availability.

In recent years, the partnership has pushed forward on development of strategies and formulation of enabling-framework conditions to scale-up dissemination. However, to move forward, there is a need to address the following:

- **Framework providing a holistic energy policy for the MSME sector**: While there is a policy-level commitment for renewable energy in general, and for integration of renewable energy and energy efficiency in large industries, the MSME sector itself does not have a holistic energy policy. Other energy-intensive sectors that have developed some frameworks for energy conservation, including an increase in the share of renewable energy, have given impetus to these sectors. A similar framework or policy/regulation is needed for the MSME sector as a whole. Such a framework could have a target in terms of increasing the share of renewable energy or introducing regulations that mandate the use of renewable energy as the main source of energy for certain levels of thermal energy use in MSMEs.
- **Strengthening and mobilization of stakeholders**: An enabling framework will assist in mobilizing industry associations to shift toward renewable-energy solutions and industries' district centers, which play a key role in supporting MSMEs to provide awareness and demonstrations of these solutions. These associations and district centers can play a critical role in strengthening institutional structures to facilitate gasifier deployment. Technical institutions are

needed for skills development and training. Similarly, the financial institutions need to develop innovative financial products for MSME clusters.

- **Cluster service approach**: In order to scale-up, there is a need to undertake a cluster approach with the objective of providing renewable-energy solutions directly or integrated with energy-efficiency measures. A cluster service approach with involvement of a local manufacturer/local service provider, which offers multiple services of technology supply, biomass supply, maintenance, and awareness creation, can be a potential model for accelerated diffusion.

In *conclusion*, to enable leapfrogging of biomass energy technologies in the MSME sector, it is important to move the solutions from research and technology demonstration (supported by bilateral agencies and the government) toward commercialization with involvement of the private sector and industries, supported by regulatory and policy frameworks. This will result in economies of scale and technology upgrades to compete with other energy or renewable-energy options.

## References

Government of India (GoI). (2008). National action plan on climate change (NAPCC). http://www.indiawaterportal.org/sites/indiawaterportal.org/files/National%20Action%20Plan%20on%20Climate%20Change%20(NAPCC)_Prime%20Ministers%20Council%20on%20Climate%20Change_Government%20of%20India_%202008.pdf. Accessed 15 March 2015.

Kishore, V. V. N., & Mande, S. (Eds.). (2007). Towards cleaner technologies: A process story on biomass gasifiers for heat applications in small and micro enterprises. New Delhi: The Energy and Resources Institute (TERI) and the Swiss Agency for Development and Cooperation (SDC).

Ministry of Micro, Small and Medium Enterprises (MoMSME). (2013). *Annual report 2012-13*. New Delhi: MoMSME, Government of India.

Planning Commission, India. (2012). *Faster, more inclusive and sustainable growth: An approach to the 12th five year plan*. Working Papers, eSocialSciences. http://EconPapers.repec.org/RePEc:ess:wpaper:id:4452. Accessed 15 March 2015.

Sinha, S. (2012). In pursuit of a light bulb and a smokeless kitchen: Longitudinal analysis of the role of energy sector policies to alleviate rural energy poverty in India. *Thesis*. Enschede: University of Twente. http://dx.doi.org/10.3990/1.9789036534826. Accessed 15 March 2015.

Soni, P. (2009). Small-scale industries: Small yet significant. In S. Narain, P. Ghosh, N. C. Saxena, J. Parikh, & P. Soni (Eds.), *Climate change: Perspectives from India*. New Delhi: UNDP.

Soni, P. (2007). Global solutions meeting local needs: Climate change policy instruments for diffusion of cleaner technologies in the small-scale industries of India. *Ph.D. Thesis*. Amsterdam: Vrije Universitiet.

# Part V
# Gender-Blind Energy Technology

# Chapter 16
# Engaging with Gender and Other Social Inequalities in Renewable Energy Projects

**Bipasha Baruah and Mini Govindan**

**Abstract** The scholarship and discourse on climate change has been dominated by natural scientists. Social scientists have only recently become involved in the debate, while natural scientists have been researching the topic for much longer. Consequently, the mainstream discourse on climate change continues to be about large-scale economic instruments and complex computer models. More recently, social scientists have pointed out the limitations of techno-centric approaches and put forward alternative frameworks such as sustainable development, climate justice, human rights, and environmental ethics for conceptualizing and operationalizing the sociocultural dimensions of climate change. They have also explored and documented some of the positive and negative consequences of adopting "green" technologies to respond to the climate crisis. However, issues related to gender equity have remained under-studied even in the work of social scientists. This chapter and the three chapters that follow (Chaps. 17–19) are a modest contribution toward addressing this knowledge gap through empirical research conducted in Peru, South Sudan, and Nigeria to understand the gendered implications and outcomes of the development and expansion of renewable energy technologies. We hope that this research will highlight the need to engage more critically and proactively with gender and other social inequalities while designing and disseminating such technologies.

## 16.1 Introduction

Transitioning to a low-carbon economy has become one of the topmost global priorities of this century. Countries around the world are experimenting with ways to make their economies "greener" by developing less polluting technologies,

B. Baruah (✉)
Women's Studies, Western University, London, Canada
e-mail: bbaruah@uwo.ca

M. Govindan
The Energy and Resources Institute (TERI), New Delhi, India

 
S. Hostettler et al. (eds.), *Sustainable Access to Energy in the Global South*,
DOI 10.1007/978-3-319-20209-9_16

creating new green jobs, or by retrofitting existing sectors such as forestry, agriculture, tourism, manufacturing, waste management, construction, public transportation, and energy production (ILO 2011). The emphasis, however, has primarily been to unlock new growth engines through techno-centric approaches. While green growth has been heralded as a new panacea to combat the climate dilemma, a clear priority for social and gender inclusiveness has yet to emerge.

A gendered analysis of green growth and development strategies reveals at least two very problematic blind spots. First, women are known to have weaker access to new technologies almost everywhere in the world (see, e.g., Rosser 2005; Hafkin and Huyer 2006) so there are likely to be unequal access issues inherent in the transition to low-carbon economies. New technologies are known to have impacts not just on individual women's lives but also on gender relations. Within the renewable energy sector, for example, there is tremendous enthusiasm about "leapfrogging" into off-grid solutions for poor rural and urban populations that are unserved or underserved by the grid, but the gender implications of such initiatives, as Álvaro Fernández-Baldor et al. conclude from their research in Peru, remain unquestioned and unexamined (Chap. 17). Several studies on off-grid solutions for electrification conducted in different world regional contexts have already highlighted the gender-differentiated impacts at the individual, household, and community levels, but the lessons of the past are not always carried forward. Gender inclusiveness in the renewable energy sector, even at the conceptual level, remains nascent almost everywhere in the world.

Second, it is well established that women are already very poorly represented globally in sectors like construction, renewable energy, manufacturing, and public transportation that are critical to the green economy. Women account for 9 % of the global workforce in construction, 12 % in engineering, 15 % in financial and business services, and 24 % in manufacturing. Women have also long been marginalized in the energy sector work force. They rarely hold ministerial positions in the field of science, technology, natural resources, and energy, and are seldom considered key stakeholders for energy initiatives. Women make up less than 6 % of technical staff and below 1 % of top managers (UN Women 2012). For a variety of reasons, globally very few women pursue the types of technical training that provide the necessary skills for contributing to innovation and technology development. Gender inequity, thus, prevents women from fully optimizing their economic potential and restricts their ability to contribute to the shift toward a greener economy.

If issues of gender equity are not addressed proactively and systematically, the green economy may do what the Green Revolution did in the 1970s—boost economic productivity by putting capital and technology in the hands of wealthier, predominantly male, farmers while marginalizing and making women in the agricultural sector even more invisible and vulnerable to poverty than they already were. In the absence of appropriately targeted multidimensional approaches such as training, education, apprenticeships, employment placement, financial tools, and supportive social policies, the green economy may exacerbate existing gender inequities and hinder poverty alleviation goals. Gender-blind approaches to economic development and environmental protection—regardless of whether they are driven by governments, civil society, private corporations, or international aid—have tended to marginalize women in their roles as workers, consumers, and citizens. For the green economy to be socially inclusive, gender equity in the broader context of social transformation and changes in production and consumption patterns must be addressed.

Although it is widely accepted that application of end-use energy technologies can improve productivity and contribute significantly to women's welfare, it does not guarantee that technological innovations can by themselves promote gender equality and women's empowerment. This is exemplified by the failure of innumerable projects promoting solar and biomass technologies to address social justice and equity concerns. This has led several researchers (see, e.g., Skutch 2005) to emphasize that the process by which the energy service is planned, implemented, and maintained, if done in a gender-sensitive manner, may be more empowering than the energy technology itself.

The papers that follow describe how market-based technocratic development "solutions" often end up entrenching rather than subverting social inequalities based on gender, class, race, and ethnicity. Samer Abdelnour (Chap. 18) and Temilade Sesan (Chap. 19) present research from south Sudan and Nigeria, respectively, to demonstrate how the need to understand complex crises is often replaced by the quest to design, produce, promote, and deliver the most profitable technologies. Both authors emphasize that doing so increases the burden of poverty and social powerlessness and silences the voices of those that the development sector is supposed to champion and serve.

We hope that these papers will encourage more critical engagement with the technologies that drive the renewable energy sector and highlight the need to engage with social inequalities while designing and disseminating them. We encourage the development community to be vigilant about the fact that even well-intentioned technical innovations that do have the potential to challenge or subvert gender inequality and other social hierarchies may end up being coopted to serve neoliberal economic goals that have nothing to do with alleviating poverty or promoting social justice.

## References

Hafkin, N., & Huyer, S. (2006). *Cinderella or cyberella? Empowering women in the knowledge society*. Boulder: Kumarian Press.

International Labour Organisation [ILO]. (2011). *Promoting decent work in a green economy*. Geneva: ILO.

Rosser, S. (2005). Women and technology through the lens of feminist theories. *Frontiers A Journal of Women's Studies*, *26*(1):1–23.
Skutch, M. M. (2005). Gender analysis for energy projects and programmes. *Energy for Sustainable Development, 9*(1), 37–52.
UN Women (2012). Fast-forwarding women's leadership in the green economy. http://www.unwomen.org/2012/06/fast-forwarding-womens-leadership-in-the-green-economy/. Accessed November 19, 2014.

# Chapter 17
# Gender, Energy, and Inequalities: A Capabilities Approach Analysis of Renewable Electrification Projects in Peru

**Álvaro Fernández-Baldor, Pau Lillo and Alejandra Boni**

**Abstract** Technology, despite being very important, is not the only factor that ensures the success of an intervention. There are many different elements to take into account when planning projects, especially in complex environments such as the least-developed areas of the world. However, development aid interventions have generally been focused on supplying a technological goods or services instead of focusing on people, thus missing out on the project's potential for social transformation. This paper analyzes four renewable energy-based electrification projects implemented by the nongovernmental organization Practical Action in the rural area of Cajamarca, Peru. Using the Capabilities Approach, the research examines the effect of the projects on the things people value. It confirms that projects provide different benefits to the communities (reducing air pollution caused by candles and kerosene, improving access to communication through television and radio, providing the possibility of night study under appropriate light, etc.), but also detects an expansion of the capabilities in other areas not considered by the nongovernmental organization such as those related to religion, leisure, or community participation. However, the expansion of capabilities is different for men and women. The study reveals the limitations of interventions designed to supply technology, electrification in this particular case, which do not take into account certain elements that can cause the use of technology to contribute unequally to the expansion of people's capabilities. The research concludes that technological projects can generate inequalities and some recommendations are presented in order to address these issues when planning interventions.

Á. Fernández-Baldor (✉) · P. Lillo
Universitat Politècnica de València, València, Spain
e-mail: alferma2@upv.es

A. Boni
Instituto de gestión de la innovación y del conocimiento (INGENIO),
Agencia Estatal Consejo Superior de Investigaciones Científicas (CSIC),
Universitat Politècnica de València, València, Spain



S. Hostettler et al. (eds.), *Sustainable Access to Energy in the Global South*,
DOI 10.1007/978-3-319-20209-9_17

## 17.1 Introduction

In rural areas, where development indexes are normally low, providing access to electricity is considered a key factor in reducing poverty (Gomez and Silveira 2010; Pasternak 2000; Borges da Cunha et al. 2007). However, the degree of isolation of non-electrified rural communities in Peru represents a significant barrier to accessing this service due to high investment costs, limited capacities of operation and maintenance (ESMAP 2001), and low consumption rates (de Gouvello 2002). In this context, small-scale off-grid renewable energy systems represent a suitable alternative for providing electricity to the rural population (for example, Nguyen 2007; Breyer et al. 2009; Benecke 2008; Chaurey et al. 2004; Lhendup 2008; Terrapon-Pfaff et al. 2014a, b). In fact, technical studies show these decentralized models can be more appropriate for rural contexts than grid extension due to the avoidance of costly and inefficient transmission losses (Benecke 2008), the suitability for low load factor projects (Kaundinya et al. 2009), and the provision of energy independence for the users (Hiremath et al. 2009; Akorede et al. 2010). Nevertheless, there are not many studies analyzing the impact of electrification in people's life.

With the aim to provide rural communities access to sustainable renewable energy-based off-grid electrification services, the international technical cooperation NGO, Practical Action has been developing energy projects in the Peruvian region of Cajamarca since 1985. As observed in previous research (Fernández-Baldor et al. 2014), a weakness in the work of Practical Action in Peru is that it focuses more on the implementation and management of the asset or technological service than on people. On the one hand, the technologies are appropriate, but the projects do not try to empower people during the process, missing their transforming potential. On the other hand, Practical Action considers the household to be its focus of intervention but does not try to understand relationships within the household. Therefore, it frequently fails to consider intra-household inequalities caused by the asset or technological service. This is particularly relevant regarding gender inequalities within families. As Baruah (2015) states, gender-blind approaches can marginalize women in their roles as workers, consumers, and citizens.

The purpose of this paper is to focus on gender inequalities caused by the electrification projects. The research uses the human capabilities approach to provide information about the implications of the use of technology vis-à-vis the real options available to the beneficiaries of electrification projects.

## 17.2 The Capabilities Approach as a Framework to Analyze Gender Inequalities

Capabilities refer to the various combinations of functionings that a person can achieve. The functionings are the different states or activities that a person values being or doing (Sen 1999). The fundamental difference between capabilities and functionings is that capabilities represent the full range of possible (achievable) functionings a person can choose from. In other words, a person can have certain capabilities (enjoy different freedoms) but choose to use them or not. These choices depend on the individual's context, personality, life history, and other factors (Robeyns 2005). The approach is people-oriented, since it recognizes that people have different personal and social factors that interfere with converting the same resource into a real option. Therefore, a main goal of the Capabilities Approach (CA) is to deal with the conversion of commodity characteristics into functionings, taking into account the attributes of individuals (intelligence, metabolism, etc.) and societies (gender roles, institutions, etc.) (Robeyns 2000).

Sen states that the focus of evaluations and judgements should be the real choices people have (capabilities), and not only the resources available. This is very relevant when analyzing gender inequalities: "the question of gender inequality… can be understood much better by comparing those things that intrinsically matter (such as functionings and capabilities), rather than just the means [to achieve them] like… resources" (1992, p. 125). Gasper and Van Staveren (2003) provide an example: A woman can be free to qualify for a public post, but it is possible that her commitments and family care responsibilities prevent her from doing so. Therefore, it is not a real option to her. In that case, the challenge is for men to assume domestic chores to create a real possibility for women. Baruah (2015), in her research about the renewable energy sector in India, finds that men are responsible for making purchasing decisions, so success in introducing a new technology in the household will depend more on men's perception of its benefits than in the actual benefits for women. Such details do matter when assessing deprivations and shaping policies that can remove particular sorts of unfreedoms.[1]

Many scholars have used the CA to address gender issues (see for example Robeyns 2003, 2008). The main advantage is the focus on individuals: Capabilities and functionings are properties of individuals, so the units of normative judgement are individuals, not households, or communities. This point is important for analyzing gender inequalities since women's well-being cannot be subsumed under wider entities. Another advantage is the inclusion of non-market dimensions of well-being, such as care labor or household work. Women are spending much more time outside the market economy than men, so the inclusion of these aspects in our normative analysis will reveal complexities in the distribution of well-being that an analysis of income alone cannot capture. The last advantage is that the approach acknowledges human diversity: The conversion of commodities into functionings can differ between people. Thus, by conceptualizing gender inequality in the space of functionings and capabilities, there is more scope to account for human diversity, including gender.

However, the main weakness of the CA in analyzing gender inequalities is its underspecified nature. As Robeyns (2008) stresses, additional theories are needed, and the selection of these theories can influence the outcome. From her point of

[1]Sen uses the term *unfreedom* for lack of freedom. See for example Sen (1999).

view, different capability assessments can be made depending on the supplementary theory used. Moreover, and according to Koggel (2013), the CA does not delve sufficiently deeply to question the actual relationships assumed by and embedded in institutions and structures or how these can entrench inequalities at various levels that are difficult to understand and to remove.

Iversen (2003) points out that power inequality in the household affects the opportunities of the family members to achieve well-being and may even distort their preferences. The same author mentions that in traditional societies, women may sacrifice their notion of well-being for the sake of the household. This point, known as adaptive preferences, is seen by many scholars as Sen's main argument against utilitarianism (see, for example, Teschl and Comim 2005; Qizilbash 2006; Robeyns 2008; Clark 2012). Sen (1985, 1992) criticizes approaches that measure well-being in terms of utility. According to Sen, unfavorable circumstances can make allies out of the disadvantaged and deprived in so far as they "learn to take pleasure in small mercies and cut down their desires to modest—'realistic' proportions in order to avoid bitter disappointment" (1992, p. 55). For Clark (2012), adapting to what is feasible may be a good thing in as much as it reduces the amount of suffering and misery associated with objectively straitened circumstances. For Sen, the propensity to minimize difficulties and enjoy small breaks "is one way of being able to live peacefully with persistent deprivation" (2009, p. 283).

For Nussbaum, unequal social and political circumstances give women unequal human capabilities: "Burdened, often, with the 'double day' of taxing employment and full responsibility for housework and child care, they lack opportunities for play and the cultivation of their imaginative and cognitive faculties" (2000, p. 220). Thus, she suggests that they may be under considerable social pressure to say they are satisfied despite the lack of these things. If people learn to adapt to their limited opportunities, for Sen it would be "ethically deeply mistaken to attach a correspondingly small value to the loss of well-being because of this survival strategy," and he suggests "the metric of happiness may, therefore, distort the extent of deprivation in a specific and biased way" (Sen 1987, p. 45). The CA insists also that we need to scrutinize the context in which economic production and social interactions take place, and whether the circumstances in which people choose from their opportunity sets are enabling and just. This context is relevant to analyzing women preferences in rural areas of Peru, because this group is particularly sensitive to the patriarchal context, lack of access to educational opportunities, and so forth.

Consequently, the concept of capabilities enables us to analyze the situation and position of women from the angle of their levels of objective well-being. And this is particularly relevant for analyzing gender inequalities in technology access or caused by technology.

In the following section, we briefly describe the field methodology implemented to operationalize the approach, but also the context in which electrification projects were developed.

## 17.3 Off-Grid Electrification Projects in Cajamarca

Cajamarca is a rural department of Peru with a low human development index (0.37) (UNDP 2013). The percentage of population with health posts' deficit is 54.6 %; the malnutrition rate is 40.9 %; and the illiteracy rate is more concentrated in women (74.6 %) than in men (25.4 %). In the sociocultural aspect, Cajamarca has a traditionally patriarchal society, illustrated by the high rates of domestic violence and female illiteracy, for instance. Underemployment and unemployment rates are high among women, and informal trade is mostly made up of women dedicated to selling agricultural surpluses, as well as sweets, cigarettes, food stands, and flowers, among other products. The migration to cities is a remarkable trend in recent years as a way to escape from poverty in rural and isolated communities. However, men have greater mobility since they have more economic resources and education, and more independence to take decisions (Arana 2012).

Practical Action operates in this context. Regarding the management protocol of the NGO, Practical Action gives priority to interventions in communities that have demanded the implementation of a project. In other cases, Practical Action selects a community and proposes a project to them. Once the project has been accepted, a socioeconomical study is undertaken, using questionnaires and interviews to community leaders. Then, the technical team prepares the technical design of the systems and the project report.

Practical Action has been developing a management scheme for the implementation of the electrification projects based on the following stakeholders: the microenterprise (run by community members, in charge of the operation, maintenance, and administration of the system); the users (the families which consume electricity); and the control unit (consisting of community people, in charge of supervising the management of the microenterprise and of dealing with the users' complaints and suggestions). Of particular relevance in this management scheme are the roles of the operator and the administrator of the microenterprise. They are in charge of the operation and maintenance as well as the collection of a monthly fee, which is saved in a joint bank account in order to replace old equipment. Between four and eight people within the community are trained to handle the equipment, and two of them are chosen in the community assembly to take these responsibilities (more information about Practical Action's management can be found in Lillo et al. 2015; Fernández-Baldor et al. 2014; Ferrer-Martí et al. 2012; Sánchez 2007).

Four electrification projects were analyzed during fieldwork. The general characteristics of the projects are shown in Table 17.1.

A few tools were introduced during fieldwork to obtain data related to the context and to the conversion factors affecting the capabilities of the people. It also focused on finding out the relationship between the electrification projects and expansion of capabilities, particularly with reference to gender issues. Another important aspect to be highlighted is the participatory character of the methodology. Table 17.2 displays the techniques used and their objectives.

**Table 17.1** Electrification projects analyzed in Cajamarca (Peru)

| Community | Technology | Installed power | Beneficiary households |
|---|---|---|---|
| Alto Perú | Microwind network | 2 kW | 11 |
| Chorro Blanco | Microhydro plant | 20 kW | 37 |
| Campo Alegre | Individual solar and wind energy hybrid | 20 × 0.15 kW | 20 |
| El Regalado | Microhydro plant | 12 kW | 31 |

**Table 17.2** Communities fieldwork

| Methodology | Technique | Data to gather |
|---|---|---|
| Focus group | Open questions on the electrification project and the community organization | Social conversion factors; analysis of the goods and services provided by the project |
| | Timeline | Community history; project relevance for the community |
| Participatory workshop | Women's workshops and men's workshops | Things people value; relationship between the project and the things people value |
| | Ball game[a] | Find out the effects of the electrification project |
| Individual interviews | Semi-structured questionnaire | People's life history; focus on individual capabilities; find out relevant personal data related to the project process (participation, equity, access, organization, etc.); find out agency related data (motivations, negotiation, participation in important decision-making, sense of community, etc.) |
| | Time use | Women's chores and men's chores (before and after the project) |

[a]The Ball game is a facilitation tool to promote participation. Standing in a circle, each participant is asked to throw a ball to another participant and to answer a question (in this case related to the impacts of electrification projects)

The methodology has an important gender component. First, the participatory workshops took place separately in each community, one for women and another one for men. This allowed women to express themselves freely without the presence of their husbands. Second, the time use technique was applied during the individual interviews to monitor the differences between women's and men's chores. In addition, the work of the focus groups provided information about women's access to posts of responsibility in the community, as well as their access to, and participation in, the electrification projects.

The participatory workshops in the communities were the key tool for the collection of information regarding the things women value and their relationship to the project. The main question discussed by the participants during the first meeting was: *what are the things or opportunities you would like your children to enjoy in the future*? The objective of asking this question was to determine the things people value in their lives. The second part of the workshop focused on the positive and negative effects of the electrification project. The participants were asked to complete the following sentence: "*I like the project because now…*" and "*I don't like the project because now…*" Then, links were established between the things people valued and the effects of the project. The workshop ended with a group reflection on how the project impacts on the things people valued.

## 17.4 Results: Gender Inequalities in Electrification Projects

According to Practical Action (Sánchez 2007), the electrification projects provide different benefits to the communities: reduced air pollution as they substitute candles or kerosene with electricity; improved access to communication through television and radio; created or improved the possibility of night study under appropriate light; the use of computers and audiovisual equipment in schools; and improvement of local medical center equipment. There is also some cost-saving for people because the cost of electricity is lower than that of candles, kerosene, and batteries. Finally, the productive use of energy in local businesses such as restaurants, hostels, and mills can improve production and sales and can thus provide an economic benefit for the community. The fieldwork research confirms those benefits but also, using the CA, detects an expansion of the capabilities in other areas not considered by Practical Action. Some participants from the Campo Alegre community highlighted the fact that the project enabled them to establish connections with other people, to read the bible at nighttime or to discover new professions through television. In the Alto Peru community, the participants valued the availability of light for their own security at night against robbery as well as the opportunity to celebrate night assemblies, which increased community participation. In the Chorro Blanco community, the participants emphasized that the light made their community more attractive, reducing the emigration of the young as well as enabling people from other communities to settle in the area. Also, in El Regalado, the participants felt that electric lighting had promoted an increased sense of collective dignity: "*We are not envious of city life*" was a statement in one of the workshops.

Nevertheless, technological projects can generate inequalities, and this has to be taken into consideration when planning this type of intervention. On the one hand, the fact that not all families enjoy electric lighting can provoke some discord in the community. While the beneficiary families acknowledge the gains made because of the project, those without the service were left behind. On the other hand, it was easy to find out gender inequalities within families.

In this paper, we will focus on the gender inequalities caused by the projects, revealed by using the CA.[2] The presence of light in homes increases the number of

[2]A broader analysis of inequalities can be found in Fernández-Baldor et al. (2014).

activities that can be performed throughout the day. However, the type of activities differs greatly between the genders, and women are not as free as men to broaden the set of capabilities regarding dimensions they value, such as leisure, religion, or community participation.

It is commonplace for men to expand their leisure time watching television or playing an instrument at nighttime, while women use this time to knit or sew until late. Although interviewed women affirmed their happiness in being able to complete these chores, in the four workshops carried out in the communities, they put knitting activities under the dimension of income generation, not under the leisure dimension. While men are free to carry out activities for personal enjoyment, women sacrifice their leisure time to increase family income. Hence, can this happiness in being able to increase their time to knit or sew be understood as an adaptive preference for women? Using the CA allows us to confirm it is.

Regarding religion, men read the bible and watch religious movies at nighttime, enjoying their spiritual development. However, women are not able to take advantage of this in the same way because the high level of illiteracy among women prevents them from reading the bible and, as stated before, the increase in their domestic night chores does not leave them time for the development of their spiritual life.

In terms of participation, women rarely attend community assemblies and, when they do, they do not generally feel as confident and as free to express their opinions as men.

Therefore, the analysis of the well-being of women in terms of capabilities and functionings reveals that men are freer than women in pursuing things they value (e.g., studying at night or participating in community assemblies). As Iversen (2003) points out, women sacrifice their own personal well-being for the sake of the household. For Nussbaum (2000, p. 220), this is a problem of social justice: "*they* [women] *lack opportunities for play and the cultivation of their imaginative and cognitive faculties*."

## 17.5 Contributions of the Capabilities Approach to Practical Action Project Planning

The following are some of the recommendations proposed to Practical Action, which were discussed and produced together with the people responsible for the energy program in Cajamarca, their technicians, and also staff from other NGOs linked to the organization. Our intention was to offer realistic recommendations, relevant to the context in Cajamarca and feasible for the organization to put into practice, disregarding options related to gender that need long-term interventions to be successful, because these energy projects are implemented within 12 and 18 months.

The first suggestion to Practical Action is to include a different type of information when planning projects. According to a Practical Action officer in Peru, the organization currently follows World Bank recommendations and only gathers socioeconomic quantitative information. This has proved to be insufficient, and it would be appropriate to gather information about values, customs, gender roles, and leadership in the community, among others.

It is also necessary to increase women's participation in all stages of the project cycle:

- First, diagnosis and technological choice are carried out by Practical Action, selecting individual or collective technological alternatives based on their own criteria. However, a complete gender diagnosis should be carried out prior to the project implementation, and gender strategies must be specifically defined for each context. Baruah (2015) shows an example in India about the importance of women participating in the design of the technology: "[…] the need for child safety devices—to enable end-users to use cookstoves safely in their homes—was identified by a female member of [the] design team who observed the difficulties women faced in cooking with small children in close proximity."
- Second, a local labor force is required for the construction of infrastructure and the installation of the systems. However, the role of women in these processes is limited to transporting materials and preparing food for men who work. Hence, initial training focused on strengthening women's technical skills could allow them to be involved in higher responsibility tasks, thus visibilizing women by enabling them to be in charge of the same activities as men.
- Third, we have identified low levels of women's attendance at community assemblies due to timetable incompatibilities with work duties, as well as passive participation behaviors because the power imbalance restricts them from publicly expressing their personal, political, or ideological opinions. To address the first issue, Practical Action should take into account the information gathered in the initial diagnosis regarding men and women's time availability and define special workshops with the community at the first stages of the project to establish the timings of the meetings. Regarding the second issue, separate meetings or workshops for men and women should be organized to ensure women's participation. Moreover, special training programs should be implemented to address women's specific needs identified in the gender diagnosis. Special attention must be drawn to reinforcing women's self-esteem and improving their argumentation and public discourse skills, so as to improve their active participation in public spaces. Furthermore, it would be recommendable to train men regarding their concept of masculinity and how it is impacting on women's lives.

Complementarily, inclusion of women in posts of responsibility should be promoted. One option could be to establish a quota of women's participation in the microenterprise and supervisory board, which will make them visible and ensure their representation, as well as improve gender equity. Another option, which could transform power structures, would be to impose that either the operator or the administrator were a woman. This would help the community realize that women can perform both in technical and administration posts, thus improving women's agency. The same method could be applied to the construction of infrastructure or the installation of energy systems where women could also become visible and perform the same tasks as men.

Finally, it would also be good to have specific gender indicators to assess the projects' success in this respect. To verify the achievement of the objectives of the project, a local monitoring committee should be defined to follow up the progress of the project. Again, women should be responsible for this task to promote agency strengthening.

Although most of these recommendations will slightly increase the costs of these kinds of projects in rural areas, an allocated budget to ensure the availability of gender resources during the project will improve the impact of these projects on women's lives to a great extent.

## 17.6 Conclusions

The research has revealed the limitations of interventions focused solely on supplying technology, electrification in this particular case, without taking into account certain societal elements that can make the use of technology contribute unequally to the expansion of people's capabilities. In the projects analyzed in this research, the most relevant factor explaining these inequalities is gender, which needs to be urgently addressed by Practical Action when planning their projects. Thus, most of the recommendations of the previous section are aimed at helping Practical Action promote a more even expansion of people's capabilities.

This study also confirms the potential of the CA by providing information about the implications of the use of and access to technology in relation to the real freedoms people can enjoy. As this study shows, this is particularly relevant when analyzing gender inequalities.

This research also has certain limitations. It has not analyzed, for instance, the development of children's capabilities due to limitations of time and resource. Likewise, the case studies could be extended to other communities and other types of technology. Research currently in progress is taking into consideration these limitations, trying to look in depth into the aspects missing in this paper.

## References

Akorede, M. F., Hizam, H., & Pouresmaeil, E. (2010). Distributed energy resources and benefits to the environment. *Renewable and Sustainable Energy Reviews, 14*(2), 724–734.

Arana, M. T. (2012). Representaciones Sociales de las Mujeres Cajamarquinas sobre la Contaminación Medioambiental Minera. *Master dissertation, Universidad Peruana Cayetano Heredia,* Peru.

Baruah, B. (2015). Creating opportunities for women in the renewable energy sector: Findings from India. *Feminist Economist,*. doi:10.1080/13545701.2014.990912.

Benecke, G. E. (2008). Success factors for the effective implementation of renewable energy options for rural electrification in India—Potentials of the clean development mechanism. *International Journal of Energy Research, 32*(12), 1066–1079. doi:10.1002/er.1445.

Borges da Cunha, K., Walter, A., & Rei, F. (2007). CDM implementation in Brazil's rural and isolated regions: The Amazonian case. *Climatic Change, 84*(1), 111–129.
Breyer, C. H., Gerlach, A., Hlusiak, M., Peters, C., Adelmann, P., Winiecki, J., et al. (2009). Electrifying the poor: Highly economic off-grid PV systems in Ethiopia—A basis for sustainable rural development. In *24th European Photovoltaic Solar Energy Conference*.
Chaurey, A., Ranganathan, M., & Mohanty, P. (2004). Electricity access for geographically disadvantaged rural communities—Technology and policy insights. *Energy Policy, 32*(15), 1693–1705.
Clark, D. A. (2012). Adaptation and development—Issues, evidence and policy relevance. In D. A. Clark (Ed.), *Adaptation, poverty and development: The dynamics of subjective well-being* (pp. 1–34). Basingstoke: Palgrave Macmillan.
de Gouvello, C. (2002). The limitations of the conventional grid: Bank on complementarity. In C. de Gouvello & Y. Maigne (Eds.), *Decentralised rural electrification: An opportunity for mankind, techniques for the planet* (pp. 127–138). Paris: Systemes Solaires.
Energy Sector Management Programme [ESMAP] (2001). *Peru: Rural electrification.* Washington D.C.: The World Bank.
Fernández-Baldor, A., Boni, A., Lillo, P., & Hueso, A. (2014). Are technological projects reducing social inequalities and improving people's well-being? A capability approach analysis of renewable energy-based electrification projects in Cajamarca, Peru. *Journal of Human Development and Capabilities, 15*(1), 13–27.
Ferrer-Martí, L., Garwood, A., Chiroque, J., Ramirez, B., Marcelo, O., Garfi, M. M., et al. (2012). Evaluating and comparing three community small-scale wind electrification projects. *Renewable and Sustainable Energy Reviews, 16*(7), 5379–5390.
Gasper, D., & van Staveren, I. (2003). Development as freedom—And as what else? *Feminist Economist, 9*(2–3), 137–161.
Gomez, M., & Silveira, S. (2010). Rural electrification of the Brazilian Amazon—Achievements and lessons. *Energy Policy, 38*(10), 6251–6260.
Hiremath, R., Kumar, B., Balachandra, P., Ravindranath, N. H., & Raghunandan, B. (2009). Decentralised renewable energy: Scope, relevance and applications in the Indian context. *Energy for Sustainable Development, 13*(1), 4–10.
Iversen, V. (2003). Intra-household inequality: A challenge for the capability approach? *Feminist Economist, 9*(2–3), 93–115.
Kaundinya, D., Balachandra, P., & Ravindranath, N. H. (2009). Grid-connected versus stand-alone energy systems for decentralized power—A review of literature. *Renewable and Sustainable Energy Reviews, 13*(8), 2041–2050.
Koggel, C. M. (2013). A critical analysis of recent work on empowerment: Implications for gender. *Journal of Global Ethics, 9*(3), 263–275.
Lhendup, T. (2008). Rural electrification in Bhutan and a methodology for evaluation of distributed generation system as an alternative option for rural electrification. *Energy for Sustainable Development, 12*(3), 13–24.
Lillo, P., Ferrer-Martí, L., Boni, A., & Fernández-Baldor, A. (2015). Assessing management models for off-grid renewable energy electrification projects using the human development approach: Case study in Peru. *Energy for Sustainable Development, 25*, 17–26.
Nguyen, K. Q. (2007). Alternatives to grid extension for rural electrification: Decentralized renewable energy technologies in Vietnam. *Energy Policy, 35*(4), 2579–2589.
Nussbaum, M. C. (2000). Women's capabilities and social justice. *Journal of Human Development, 1*(2), 219–247.
Pasternak, A. D. (2000). *Global energy futures and human development: A framework for analysis*. Livermore: Lawrence Livemore National Library.
Qizilbash, M. (2006). Well-being, adaptation and human limitations. *Royal Institute of Philosophy Supplements, 59*, 83–110.
Robeyns, I. (2000). An unworkable idea or a promising alternative? Sen's capability approach re-examined. In *Center for economic studies—Discussion paper ces0030*. Leuven: Katholieke Universiteit.

Robeyns, I. (2003). Sen's capability approach and gender inequality: Selecting relevant capabilities. *Feminist Economics, 9*(2–3), 61–92.
Robeyns, I. (2005). The capability approach: A theoretical survey. *Journal of Human Development, 6*(1), 93–117.
Robeyns, I. (2008). Sen's capability approach and feminist concerns. In M. Comim, M. Qizilbash & S. Alkire (Eds.), *The capability approach: Concepts, measures and applications* (pp. 82–104). Cambridge: Cambridge University Press.
Sánchez, T. (2007). *Organización de servicios eléctricos en poblaciones rurales aisladas*. Lima: Soluciones Prácticas—ITDG.
Sen, A. (1985). Well-being, agency and freedom: The Dewey lectures 1984. *The Journal of Philosophy, 82*, 169–221.
Sen, A. (1987). *On ethics and economics*. Oxford and New York: Basil Blackwell.
Sen, A. (1992). *Inequality reexamined*. New York, Oxford: Russell Sage Foundation, Clarendon Press.
Sen, A. (1999). *Development as freedom*. New York: Oxford University Press.
Sen, A. (2009). *The idea of justice*. London: Allen Lane.
Terrapon-Pfaff, J., Dienst, C., König, J., & Ortiz, W. (2014a). How effective are small-scale energy interventions in developing countries? Results from a post-evaluation on project-level. *Applied Energy, 135*, 809–814.
Terrapon-Pfaff, J., Dienst, C., König, J., & Ortiz, W. (2014b). A cross-sectional review: Impacts and sustainability of small-scale renewable energy projects in developing countries. *Renewable and Sustainable Energy Reviews, 40*, 1–10.
Teschl, M., & Comim, F. (2005). Adaptive preferences and capabilities: Some preliminary conceptual explorations. *Review of Social Economy, 63*(2), 229–247.
United Nations Development Programme [UNDP] (2013). Informe sobre Desarrollo Humano Perú. (2013). *Cambio climático y territorio: Desafíos y respuestas para un futuro sostenible*. Lima, Perú: Programa de las Naciones Unidas para el Desarrollo.

# Chapter 18
# The Cookstove–Rape Prevention Myth and the Limits of Techno-saviorism

**Samer Abdelnour**

**Abstract** The field of humanitarian technology is rife with *techno-saviorism*—the promotion of technology as a panacea and not simply a technique of intervention. Humanitarian technology advocates, designers, and promoters uphold technological objects and interventions as having an incredible agency to solve complex social and environmental problems. A prevalent example is the promotion of improved cookstoves to address a number of complex global problems such as the negative health impacts of cooking, climate change, deforestation, and sexual violence in conflict zones. In this paper, I outline research that critiques the "stoves reduce rape" rhetoric, and discuss techno-saviorism in terms of improved cookstove myths, consequences, and gender inequalities. I conclude with reflections from discussions with scientists and humanitarian practitioners who design, test, market, deliver, and advocate for improved cookstoves.

## 18.1 Introduction

*Technology* (with a capital *T*) has long been integral to economic development policy, given its importance to the goals of economic growth, industrialization, and Schumpeterian notions of innovation (Escobar 1995). In contrast, *intermediate* or *appropriate technology* (lowercase *t*) emerged as a movement to advance small-scale technical interventions to address the everyday context-dependent problems impoverished peoples face (Schumacher 1973). Processes associated with hyper-neoliberal globalization (unprecedented individualism, scale-ism, technocracy, and interventionism) have led to the conflation of big "T" and little "t" notions of technology. The result is a conceptual blurring between the grand ideals of

---

S. Abdelnour (✉)
Rotterdam School of Management, Erasmus University, Rotterdam, The Netherlands
e-mail: abdelnour@rsm.nl

 
S. Hostettler et al. (eds.), *Sustainable Access to Energy in the Global South*,
DOI 10.1007/978-3-319-20209-9_18

technology (as a driver of innovation, economic growth, and development) and technology as a scientific object (as a product, invention, or intervention). This has led to the rise of *techno-saviorism*: the promotion of technology as a humanitarian panacea and not simply a technique of intervention. Examples are all around us: mosquito nets to stop malaria, vaccinations to ameliorate disease, microfinance as a panacea for poverty, mobile phones for socioeconomic inclusion, and social media an enabler of democratic revolution. Some technologies appear to have distinct potency in terms of their marketed ability to solve the world's most pressing problems.

Two recent books, *Techno-Fix: Why Technology Won't Save Us Or the Environment* (Huesemann and Huesemann 2011), and *To Save Everything, Click Here: The Folly of Technological Solutionism* (Morozov 2013), go a long way to illustrate the potential consequences of overstating technical solutions. However, the field of humanitarian technology has thus far escaped significant critique. For instance, when humanitarian technology interventions fail, rather than consider the technology unsuitable intervention failures are often attributed to users, implementing agencies, poor governance, corruption, or technical issues that might be designed away. The more invested scientists, advocates, donors, and humanitarian agencies are in a particular technological intervention, the less likely they are to critique its suitability for addressing the problems it is suggested to resolve. An example are improved cookstoves (or fuel-efficient stoves), which are marketed as capable of resolving problems directly attributable to cooking (such as smoke inhalation), as well as a host of complex problems that at their root appear to have very little to do with poor women cooking (such as deforestation, climate change, and sexual violence).

This paper summarizes research that critiques how improved cookstoves came to be framed a tool to reduce sexual violence in conflict and refugee camp contexts globally, without evidence to support such a claim. The material used to explicate the construction of "stoves reduce rape" is drawn from a recent publication, titled *Technologizing Humanitarian Space: Darfur Advocacy and the Rape-Stove Panacea* (Abdelnour and Saeed 2014), which was presented at the EPFL-UNESCO Conference on Technologies for Development (Tech4Dev 2014) held in Lausanne, Switzerland. In addition to highlighting policy- and practice-relevant themes related to the study, I reflect on discussions with scientists and development practitioners involved in the design, testing, marketing, and delivery of improved cookstoves.

## 18.2 Unpacking the Cookstoves–Rape Prevention Myth

Improved cookstoves were first promoted as a tool to reduce sexual violence in 2004, during the Darfur humanitarian response. The target of stove interventions were displaced women and girls, who were exposed to sexual violence during the collection of fuelwood. The explicit assumptions that hold stoves as capable of reducing conflict-related sexual violence are as follows: (i) improved cookstoves reduce the amount of fuelwood required for cooking; (ii) women are attacked while traveling to collect wood for cooking; and (iii) the reduction in fuelwood requirements will correlate with reduced travel time and reduced exposure to sexual violence. Key implicit assumptions of this scenario are as follows: (iv) women are safe from sexual assault when not collecting firewood; (v) women do not collect wood for other purposes; and (vi) women do not travel long distances for other reasons (such as collecting grasses, water, trade, or work). Taken together, these assumptions convey a powerful narrative of risk when collecting firewood motivating an immense improved cookstove effort for Darfur (Abdelnour and Branzei 2010). Though there is little evidence to suggest that improved cookstove interventions reduce sexual attacks, in 2010 the "stoves reduce rape" rhetoric was adopted by the Global Alliance for Clean Cookstoves, a US-based initiative to promote a global market for efficient stoves, along with claims that stoves reduce climate change and the negative health impacts of cooking.

Having spent time in Darfur researching the competitive dynamics among stove-promoting organizations, it was evident that the assumptions upholding cookstoves as a protection tool do not reflect the complex reality of gender violence. Realizing that techno-centric research often fails to question the very suitability of a given technology to address specific problems, my colleague Akbar Saeed and I sought to critically examine how problems are conceived as having technology solutions, and what the potential consequences of doing so are (Abdelnour and Saeed 2014). Given this interest we began with an unconventional research question. Rather than ask a question that takes the marketed benefits of stoves as given (for instance, "What design or behavioral changes will better enable improved cookstoves to reduce fuel use, hence the time required for collecting fuelwood and the risk of sexual violence?") we motivated our inquiry with: "How is the problem of sexual violence understood such that improved cookstoves are thought a viable solution?"

To answer this question, we use the concept of *problematization* (Foucault 1988), which is understood as the process whereby problems are construed to have actionable solutions. We used corpus construction procedures for selecting and structuring qualitative empirical material (Bauer and Aarts 2000), including humanitarian advocacy marketing materials, reports, and interviews, together with discourse analysis as the primary means of analysis (Fairclough 1989, 2006). Our investigation revealed three notable "discursive cases" relevant to the construction of the idea that "stoves reduce rape": Dadaab, Darfur, and global.[1]

[1]For an expanded presentation of the case study and analysis see Abdelnour and Saeed (2014).

## 18.3 The Discursive Construction of "Stoves Reduce Rape"

The story begins in the Dadaab refugee camps in the 1990s, where Refugees International first promoted the link between firewood collection and sexual attacks. At Dadaab, displaced women and girls were exposed to high levels of sexual violence in and around the camps. According to a chairwoman of an anti-rape committee at Dadaab, the problem of sexual violence was widespread and thus demanded a comprehensive security solution. Refugees International advocates, however, focused on one actionable dimension of gender violence: firewood. Based upon the rape–firewood problematization, efforts were made to provide vulnerable women with firewood as a means to reduce attacks. Table 18.1 presents representative quotes pertaining to the firewood–rape problematization.

Assessments of firewood interventions were shown to have little impact on overall incidences of violence at Dadaab. Organizations only measuring firewood-related rape were sometimes able to show reductions in (firewood-related) violence; however, those measuring sexual violence more comprehensively determined that providing women and girls with firewood had little impact on overall levels of violence. Rather, the provision of firewood shifted the place and context of violence. As one assessment noted:

> Banditry and acts of sexual violence, especially rape, were known to occur frequently in the camps. Considerable publicity highlighted the rape of women while collecting firewood outside the camps (WCWRC 2002: 73).

As the Darfur conflict became newsworthy in the USA, an impressive Darfur advocacy network emerged, much of it revolving through the activities of religious, political, and civil society organizations. The messages promoted by US-based Darfur advocates often had little to do with the reality of violence in Darfur, instead, they drew upon post-9/11 Orientalist imagery to construe "Darfur" as a place where "Arab" aggressors preyed on helpless "African" victims (Mamdani 2009). According to Hicks (2010: 267), Darfur advocates such as "SDC [Save Darfur Coalition] and Refugees International relied on multiple interwoven constructs of religion and race and a series of overlapping binaries that specified gendered norms of citizenship, simplified dynamics in Darfur into genocidal rape, and defined American pluralist practice in terms of human rights." Along with this racial dichotomy, spatial and gender frames influenced the repackaging of the firewood–violence framing for Darfur. A simple narrative of violence emerged: in Darfur, African women and girls were exposed to sexual violence at the hands of Arab aggressors, particularly when searching for fuelwood for cooking. Table 18.2 presents an overview of representative quotes pertaining to the genocidal rape–improved cookstoves problematization.

Brimming with racial and gender stereotypes, the "Arab genocidal rape" framing was quickly embraced by Darfur advocates and their constituents, who mobilized a

**Table 18.1** Dadaab, rape–firewood problematization (1990s+)

| Description | Representative quote |
|---|---|
| Quote by a chairwoman of an anti-rape committee in one of the Dadaab camps, reproduced by Refugees International | If you go out of the camp very near, there is a man standing there with a gun. If you go to run, he shoots you. If you stay, he rapes you. Some of us are taken from our houses at night. Some of us are taken in the bush when we look for firewood. We are not safe, night or day. If we don't get security, we won't survive (Fitzgerald 1998: 2) |
| Refugees International advocates, reframing the complexity of violence at Dadaab to one with an actionable solution: firewood | ***Refugees International*** **disagrees with the proposed policy of providing only "vulnerables" with firewood. All women and young girls in the refugee camps are equally vulnerable and should receive firewood. All share the same risk of being brutally attacked if they must gather firewood** (Fitzgerald 1998: 3, italics and bold original) |
| Refugees International advocates, postulating a more globalized rape–firewood connection | One is left with the sense that international agencies still do not really know how to deal with the relationship between firewood and rape. Women are 58 % of the world's adult refugee population. They do not shirk from their responsibility to gather firewood for their families. The international community should protect them (Fitzgerald and Lowman 1998: 2) |

**Table 18.2** Darfur, genocidal rape–stove problematization (2004+)

| Description | Representative quote |
|---|---|
| Refugees International advocate, in an early framing of the Darfur conflict | The fate of a raped woman in an Islamic fundamentalist society such as that in Sudan is already sealed. […] The women of Darfur report feeling betrayed by their Islamic government, which has formed an alliance with Muslim and Arab rapists rather than protecting its own civilians (Lumeya 2004) |
| Refugees International advocating for widespread promotion of improved cookstoves to prevent gender violence, including rape | By reducing the need for wood and emission of smoke, a switch to simple, more fuel-efficient stoves could reduce the time women spend collecting wood, a task that exposes them to the risk of rape and other forms of gender-based violence (Wolf 2005: 1) |
| Jewish World Watch (JWW), founded specifically to advocate on the issue of Darfur, presents the rape–stove problematization | Women and girls who have fled the genocide in Darfur, Sudan, are particularly vulnerable to rape while performing the critical task of collecting firewood for cooking. **The Solar Cooker Project's mission is to reduce the frequency of these crimes by providing refugee women with an alternative cooking option: the solar cooker**. Solar cookers enable women to reduce their dependency on wood and remain within the safety of the camp (JWW 2012: 1, bold original) |
| Potential Energy (formerly the Darfur Stoves project), in a Mother's Day fundraising e-mail | For the women in Darfur, cooking on simple stoves in crowded camps for displaced people can bring untold hardship. The daily searches for firewood expose them to attacks and sexual assault, and looking for work to earn money for fuel takes up much of their time and energy (Neichin 2011) |

tremendous amount of Darfur action in the USA. This very specific framing of the problem of violence in Darfur permitted improved cookstoves—a simple domestic technology—to be advanced as an effective tool to reduce sexual violence. Refugees International made a widely cited "call to stoves" advocating for the widespread promotion of improved cookstoves to protect Darfuri women and girls from sexual violence.

Within a relatively short time, humanitarian agencies working in Darfur recognized the lack of evidence supporting the notion that stoves reduced sexual violence. As with the case of Dadaab, the complexity of violence in Darfur was far too complex for a simple technical solution. A Refugees International report retracted the rape–stove problematization, but was careful to maintain the suggested value of improved cookstoves, which by this time had become an important humanitarian technology intervention:

> While there is little evidence that producing fuel-efficient stoves reduces violence against women [...] the international community should continue to promote them but not solely or even principally as a protection measure against sexual violence but as a vital part of a holistic response to the urgent environmental and humanitarian issues confronting the conflict-affected peoples of Darfur (Martin 2007: 18).

Ironically, though Refugees International noted the lack of evidence supporting the "stoves reduce rape" rhetoric, there was (and remains) little concrete evidence to support the notion that improved cookstoves effectively addressed the other "urgent environmental and humanitarian issues" facing Darfur's displaced populations. Another assessment points to the complex layers of violence facing displaced women in Eastern Chad, including Darfuri refugees:

> Amnesty International's recent research demonstrates that refugee women and girls in Eastern Chad continue to be raped and attacked when they venture outside refugee camps. The attacks are carried out by organized groups, bandits and even by members of the Chadian National Army. Refugee camps are not always safer. Amnesty International has also documented cases of personnel working with international humanitarian nongovernmental organizations (NGOs) operational in the refugee camps committing acts of rape and other violence against women and girls inside the camps. Additionally, refugee women and girls experience rape and other violence at the hands of their spouses, family members and other refugees within the camps (Amnesty International 2009: 11).

Despite lack of evidence, US advocacy groups and humanitarian agencies continued to promote stoves as a means to protect Darfuri women and girls. Some explanations for this are as follows. First, the prospect of preventing a horrific crime such as sexual violence through a scalable technical solution is an attractive proposition for donors and the general public alike (for instance, "donate US$40 to prevent rape in Darfur"). In fact, promoters of stoves for Darfur, including scientists and nonprofit organizations, received a great deal of resources through donor grants, awards, and public donations for the promotion of stoves. Second, the "Arab genocidal rape" discourse suggests that stove initiatives became a tool through which to express a definitive political position about the conflict in Darfur. Seen from this perspective, stoves become a political object capable of legitimating narratives of faraway conflicts through marketing campaigns (for instance, "donate

**Table 18.3** Global rape–stove problematization (2010+)

| Description | Representative quote |
|---|---|
| Women's Refugee Commission advocates advancing a global rape–stove problematization | In hundreds of refugee and IDP [Internally displaced person] settings throughout the world, women and girls are made more vulnerable to sexual violence because of the almost daily need to leave camps in search of firewood. More can and must be done to reduce the risk (Patrick 2007: 40) |
| Global Alliance for Clean Cookstoves factsheet | Women and girls face severe personal security risks as they collect fuel, especially those living in communities of instability, including refugee camps and conflict zones (U.S. Department of State 2010) |
| Potential Energy advances a general decontextualized narrative that suggests refugee camps as relatively safe and violence as happening "out there" | Women and girls must walk many hours, several times a week, just to find a single tree with usable wood to fuel their fires. Outside the relative safety of refugee camps, they are vulnerable to acts of violence (Potential Energy 2012) |
| International Rescue Committee, in a press release about their work in the Democratic Republic of the Congo, present the stove–rape problematization alongside a new framing, "clothes reduce rape" | The women here are forced to venture out of the camp to collect firewood or water. Both make them vulnerable to rape and many of the women and girls have been assaulted. All the International Rescue Committee, which runs the camp, can offer to mitigate the threat are 'dignity kits' that contain efficient stoves that require less firewood and extra clothes so the women have to look for washing water less often (Borger 2013) |

US$40 to prevent the rape of African women by Arab militiamen in Darfur"). A third explanation is the culture of techno-saviorism, which holds humanitarian technology as having an incredible agency to resolve complex social and environmental problems.

The persistence of the rape–stoves problematization among US-based stove promoters enabled a more generalized narrative to be adopted by humanitarian advocates. Through their activities, "stoves reduce rape" became a global rhetoric. Examples of the globalized (generalized, decontextualized) rape–cookstoves problematization are presented in Table 18.3.

## 18.4 Myths, Consequences, Gender Injustices

"Stoves reduce rape" is a humanitarian technology myth; improved cookstoves are deemed effective not because of empirical evidence, but from powerful narratives that claim what they are able to accomplish (Abdelnour and Saeed 2014). This critique of the rape–stoves problematization can be extended to other marketed benefits that are attributed to improved cookstoves. For instance, the investigation of causal relationships between burning fuel in a particular stove design and measurable outcomes is relatively straightforward (burning X fuel in Y stove produces Z heat, emissions, smoke, etc.). However, claims that more efficient cooking reduces complex problems (burning X fuel in Y stove reduces C complex problems) often rely on scientific narratives of what stove technologies are capable of in lieu of supporting evidence. Moreover, certain "scientific" stove indicators are often assumed to be a viable proxy for the problems stoves are reported to address. For instance, in-laboratory efficiency tests are assumed to correlate with actual use, which in turn are assumed to correlate to changes in user behaviors, which in turn are assumed to correlate with health and environmental benefits. The assumptions inherent in all aspects of stove intervention narratives—from design to testing to actual use to outcomes—are extensive. As stove promoters become increasingly dependent on the legitimacy of these narratives, the actual effectiveness (or ineffectiveness) of stove interventions is inconsequential (Abdelnour and Saeed 2014).

The construction of the rape–stoves myth emerged through the concerted actions of many humanitarian actors, including donors, advocacy groups, research institutes, government and development agencies, and nongovernmental organizations (see Abdelnour and Branzei 2010). Humanitarian marketing (for purposes of fundraising) and advocacy (political and policy marketing) helped to elevate the "stoves reduce rape" narrative to a level of policy legitimacy, even without supporting evidence.

By construing complex sociopolitical challenges as "manageable problems," humanitarian actors are fundamental in shaping humanitarian technology policy and practice. However, simple problematizations often preclude a holistic understanding of the problems user–beneficiaries face. In the case of improved cookstoves, this includes the violence that displaced women and girls face inside and outside of camps. Instead of appreciating the complexity of problems in context, the marketing of humanitarian technologies often legitimates a generalized (often global) applicability. In Darfur, this shifted the governing questions many humanitarian actors asked, from "How do we understand and address the complex problem of sexual violence?" to "How do we design, produce, promote, and deliver the most fuel-efficient stoves?" (Abdelnour and Saeed 2014).

This "technologizing" of humanitarian thinking and practice is not without consequences. Using the empowerment of women as a pretext and marketing/advocacy platform, the promotion of improved cookstoves inadvertently transfers the responsibility for resolving complex problems into the households of the global poor. For instance, in addition to sexual violence, improved cookstove promoters suggest that through more efficient cooking, poor women are empowered to address complex problems such as deforestation and climate change. Such ignores the fact that poor women and inefficient cooking are neither at the root causes of, nor the major contributors to these problems. It also negates the momentous role of global industry (including extractive industries, large-scale logging, livestock production, and transportation) as well as the consumption patterns of the haves (especially in the West) in creating and maintaining these problems.

Moreover, efforts to carbon-certify stove initiatives will place poor women and girls in a position where they unknowingly, and through sheer necessity, serve to subsidize the polluting activities of global industry. The promise of carbon finance should not be taken as a simple "win–win" strategy. Mobilizing poor women within deeply unequal neoliberal structures raises ethical questions for donors, humanitarian intermediaries, and household energy market actors.

Technology certainly has its limits, however, so does the conceptualization of complex humanitarian problems as having simple technical solutions. The concept of problematization encourages the analysis of problems and solutions as mutually construed. For instance, improved cookstoves were first promoted as capable of reducing sexual violence in Darfur because of the specific way violence in Darfur was imagined and promoted by humanitarian and political advocates. Once disconnected from the specific context of Darfur, the generalized problem–solution construction permits improved cookstoves to be a panacea for sexual violence in any conflict zone. Humanitarian accountability requires a more critical understanding of the role of advocacy and marketing in the defining of humanitarian problems, policies, and technology interventions. Evidence-based policymaking must ground the claims made by humanitarian technology advocates in the reality of user–beneficiary experiences and not in the promise of technology.

The culture of techno-saviorism discourages appropriate critique among many improved cookstove advocates. The unwillingness to fundamentally question the suitability of improved cookstoves to solve complex problems enables their repackaging as solutions to impending crises (Abdelnour 2011).

Given the inseparability of gender inequalities and household energy, the measurable benefits of improved cookstoves for poor households warrants more critical investigation and debate (Baruah 2015). It is not enough to simply market stoves through generalized, simplified, and disingenuous narratives. Stoves are not techno-superheroes; cooking with better stoves will not enable poor women to resolve the structural causes of poverty, violence or climate change.

Simplistic narratives of gender violence in conflict and low-income contexts motivates tremendous efforts in terms of advocacy, fundraising, development policy and programming. This is in stark contrast to contested techno-solutions that seek to address sexual violence in Western contexts. Examples include nail polish to detect the presence of incapacitating (date-rape) drugs in drinks, anti-rape shorts to help women resist sexual attack, and the *SlutWalks* protests in response to suggestions by police officials that women provoke assault through the clothes they wear. It is difficult to imagine a scenario where humanitarian advocates would promote "dignity kits" (which include stoves, wash buckets, and clothing) as a means to reduce sexual violence in Western contexts (see the final quote in Table 18.3); the fact that these simple domestic technologies/objects are viewed as solutions for the Global South reveals the extent to which some humanitarian technology interventions are inundated with gender and other stereotypes.

## 18.5 Concluding Remarks

In conclusion, I would like to share some reflections that have troubled me since 2006, when I first began to conduct fieldwork on improved cookstoves in Darfur. This relates to a lack of willingness on the part of many improved cookstove advocates to engage with critical assessments of their interventions. Reluctance has come in many forms, including: (i) withholding sign-off on the use of interviews after agreeing and granting access; (ii) conference sponsors requesting that names of stove-promoting organizations be removed from presentations of research; (iii) attempts by stove enthusiasts and nonprofits to discredit research through argumentum ad hominem; (iv) dismissing established social science research designs and analyses not on their merits (such as quality, rigor, and contributions), but for being "unscientific" (with the implication that such approaches are unimportant and can therefore be ignored); and finally, (v) an unwillingness to engage in constructive conversations regarding the limits of humanitarian technology, especially with respect to the complex problems they are reportedly capable of resolving.

While I appreciate that humanitarian technology proponents and critical social scientists are often invited to present work at the same venue, such as with the Tech4Dev 2014 conference, there is often a lack of dialogue between these "hard" and "soft" traditions. Sometimes this is implicit, such as the relegating of a critical topic (for instance, gender perspectives) to one panel. However, this is also part of the techno-saviorism culture, which frames problem solving as a matter of technical innovation, better design, and scale, effectively placing technology outside the realm of genuine critique. Some of this denial comes from convincing myths that tell a story of causality in the absence of supporting evidence. In some cases, myths present technological objects as techno-superheroes (for instance, "stoves reduce rape"). Social scientists have long noted that some social and environmental problems are simply too complex for a technology solution (French 1986). Forums that instigate constructive dialogue among humanitarian technology advocates, critics, and user–beneficiaries will be a welcome innovation.

## References

Abdelnour, S. (2011). Repackaging the efficient stove as the 'solution' to crises in Darfur. *Boiling point 59*. http://www.hedon.info/BP59_From+Fuelwood+to+GBV. Accessed 22 March 2015.

Abdelnour, S., & Branzei, O. (2010). Fuel-efficient stoves for Darfur: The social construction of subsistence marketplaces in post-conflict settings. *Journal of Business Research, 63*(6), 617–629.

Abdelnour, S., & Saeed, A. M. (2014). Technologizing humanitarian space: Darfur advocacy and the rape-stove panacea. *International Political Sociology, 8*(2), 145–163.

Amnesty International. (2009). *'No place for us here': Violence against refugee women in Eastern Chad*. London: Amnesty International Publications.

Baruah, B. (2015). Creating opportunities for women in the renewable energy sector: Findings from research in India. *Feminist Economics, 21*(2), 53–76.

Bauer, M. W., & Aarts, B. (2000). Corpus construction: A principle for qualitative data collection. In M. W. Bauer & G. Gaskell (Eds.), *Qualitative researching with text, image and sound: A practical handbook* (pp. 19–37). London: Sage.

Borger, J. (2013). Congo receives £180 m boost to health system to tackle warzone rape. The guardian, March 26. http://www.guardian.co.uk/world/2013/mar/26/uk-congo-tackle-warzone-rape. Accessed 22 March 2015.

Escobar, A. (1995). *Encountering development: The making and unmaking of the third world.* Princeton, NJ: Princeton University Press.

Fairclough, N. (1989). *Language and power*. London: Longman.

Fairclough, N. (2006). *Language and globalization*. London: Routledge.

Fitzgerald, M. A. (1998). *Firewood, violence against women, and hard choices in Kenya, January 6*. Washington. DC: Refugees International. http://www.refworld.org/pdfid/47a6ebaba.pdf. Accessed 22 March 2015.

Fitzgerald, M. A., & Lowman, S. (1998). *Protect refugee women as they gather firewood*. The New York Times, August 27. http://www.nytimes.com/1998/08/27/opinion/27iht-edmary.t.html. Accessed 22 March 2015.

Foucault, M. (1988). The concern for truth. In L. D. Kritzman (Ed.), *Politics, philosophy, culture: interviews and other writings* (pp. 1977–1984). New York: Routledge.

French, D. (1986). Confronting an unsolvable problem: Deforestation in Malawi. *World Development, 14*(4), 531–540.

Hicks, R. R. (2010). Saving Darfur: Enacting pluralism in terms of gender, genocide, and militarized human rights. In P. E Klassen & C. Bender (Eds.), *After pluralism: Reimagining religious engagement* (pp. 252–276). New York: Columbia University Press.

Huesemann, M., & Huesemann, J. (2011). *Techno-fix: why technology won't save us or the environment*. Gabriola Island, BC: New Society Publishers.

Jewish World Watch (JWW) (2012) *Help the women of Darfur, March*. http://www.jww.org/wp-content/uploads/2010/06/JWW_FactSheet_Apr12_v1.pdf. Accessed 22 March 2015.

Lumeya, F. (2004). *Rape, Islam, and Darfur's women refugees and war-displaced, August 24*. Washington, DC: Refugees International.

Mamdani, M. (2009). *Saviors and survivors: Darfur, politics, and the war on terror*. New York: Pantheon Books.

Martin, S. (2007). *Ending sexual violence in Darfur: An advocacy agenda*. Washington: Refugees International. https://www.refintl.org/policy/in-depth-report/ending-sexual-violence-darfur-advocacy-agenda. Accessed 22 March 2015.

Morozov, E. (2013). *To save everything, click here: The folly of technological solutionism*. New York: Public Affairs.

Neichin, G. (2011). *Honor your mother by supporting a woman in Darfur [Email Campaign], May 3*. Berkeley, CA: The Darfur Stoves Project.

Patrick, E. (2007). Sexual violence and firewood collection in Darfur. *Forced Migration Review, 27*, 40–41.

Potential Energy. (2012). *Why stoves?* http://www.potentialenergy.org/why-stoves/. Accessed August 3, 2012. (Note: website text now changed; quote can be found through Internet search).

Schuhmacher, E. (1973). *Small is beautiful: Study of economics as if people mattered*. London: Blond & Briggs.

U.S. Department of State. (2010). *The United States and the Global Alliance for Clean Cookstoves: Factsheet, September 21*. http://www.state.gov/r/pa/prs/ps/2010/09/147490.htm. Accessed 22 March 2015.

Wolf, D. (2005). *Sudan: Rapidly expand the use of fuel-efficient stoves in Darfur*. Refugees International Bulletin, October 24. www.unhcr.org/refworld/pdfid/47a6eec98.pdf. Accessed 22 March 2015.

Women's Commission for Refugee Women and Children (WCWRC) (2002). *UNHCR policy on refugee women and guidelines on their protection: An assessment of ten years of implementation*. http://www.refworld.org/pdfid/48aa83220.pdf. Accessed 22 March 2015.

# Chapter 19
# Deconstructing 'Discriminatory' Technologies: Insights into Inclusive Development from Improved Cookstove Projects in Nigeria

**Temilade Sesan**

**Abstract** Efforts to develop and disseminate improved cookstoves among biomass-reliant populations have been launched by both Northern and Southern development actors, with the interactions between North and South constantly shaping notions of appropriate technology within local contexts. This paper examines elements of the interaction between the Centre for Household Energy and the Environment (CEHEEN), a local nongovernmental organization (NGO) that set out independently to promote the uptake of improved cookstoves in Nigeria, and Project Gaia, a US-based international NGO that arrived later on the scene with similar objectives but a different approach. Using data from interviews and project documents, the paper shows how the principles that informed CEHEEN's Improved Egaga project—those of participatory development, indigenous technology, and local production—have given way to the seemingly progressive technology transfer principles underpinning Project Gaia's CleanCook project. Contrary to CEHEEN's assumption that the ethanol-fueled CleanCook stove would be less 'discriminatory' than the wood-burning improved Egaga stove, the findings show that higher-income households that already have access to comparable alternatives such as kerosene and liquid petroleum gas (LPG) stoves are likely to be incorporated in the CleanCook solution to a greater degree than the low-income biomass-reliant households that were originally targeted by the Improved Egaga project. The paper highlights the pitfalls inherent in homogenizing the energy needs and capabilities of Southern populations and concludes that a differentiated approach to collaborative energy projects which recognizes the opportunities and limitations of households at various socioeconomic levels is likely to produce more inclusive, if modestly incremental outcomes.

T. Sesan (✉)
Centre for Petroleum, Energy Economics and Law,
University of Ibadan, Ibadan, Nigeria
e-mail: temi@gbengasesan.com



S. Hostettler et al. (eds.), *Sustainable Access to Energy in the Global South*,
DOI 10.1007/978-3-319-20209-9_19

## 19.1 Introduction

About 2.8 billion people, mostly in sub-Saharan Africa and Asia, cook with solid biomass fuels (primarily fuelwood, charcoal, and coal) over open fires and rudimentary stoves, with adverse effects on their health and environment (World Bank 2013). Despite its significant oil wealth, Nigeria features prominently in this category: An estimated 67 % of its 140 million people (2006 national census figures) rely on solid biomass fuels to meet their cooking energy needs (WHO 2006a). Attempts to alleviate the energy poverty of these populations through improved cookstove initiatives have not been widespread at country level—they have certainly not occurred on a scale equivalent to the magnitude of the problem—although there have been signs of increased activity following the establishment of the United Nations Foundation-led Global Alliance for Clean Cookstoves in 2010. It therefore seemed to constitute a significant step forward when, in 2003, US-based Project Gaia launched a series of consultations with Centre for Household Energy and the Environment (CEHEEN), a local nongovernmental organization (NGO) which had been working since the late 1990s to develop and disseminate improved stoves among biomass-reliant households in the country. On the strength of these consultations, an alliance was birthed between the two organizations which came to be christened Project Gaia Nigeria. This North–South collaboration subsequently opened the way for the institution of high-level partnerships between Project Gaia and relevant departments within state and local governments in the proposed pilot region of the Niger delta (Stokes and Ebbeson 2005).

Project Gaia's proposal was the Sweden-manufactured and patented CleanCook stove powered by alcohol fuels (methanol and/or ethanol), promoted on the basis that the stove was superior in technical performance, and therefore in its ability to deliver health and environmental benefits to energy users, than wood-burning variants of improved stoves.

To test the market feasibility and acceptance of the technology in Nigeria, two CleanCook pilots were conducted in Delta state, one of nine states in the oil-rich Niger delta. The first 'mini-pilot' in 2003 placed stoves in 15 homes and monitored their use by the female heads of those households, as it is women who traditionally do the cooking in the project region. Encouraged by the overwhelmingly positive response of the women to the stoves and eager to test the acceptance on a larger scale, a second, 'full' pilot was conducted in 2007 with 150 stoves. The households selected for the full pilot were spread across urban and rural communities in nine locations within Delta state. In addition, selection attempted to be sensitive to income level: Households were drawn from among the low-, medium-, and high-income groups. This differentiation of households by income level was especially important because, as a baseline study of energy use conducted by Project Gaia Nigeria in the previous year showed, the quality of the primary cooking fuels used by households in the state is directly proportional to their income: Generally, the lower households are in the income pyramid, the lower the quality of their primary cooking fuel, and vice versa. This gradation of primary household fuel type by income level has been captured in the concept of the energy ladder (q.v. Leach 1992; Pachauri and Spreng 2003), which depicts low-income populations as being most heavily reliant on solid biomass fuels and illustrates the tendency for households to move toward more modern fuels with increasing incomes. As we will see, however, the differential energy use patterns indicated by Project Gaia's baseline study findings did not appear to inform the organization's outlook on the feasibility of uptake of the CleanCook stove and fuel by low-income households in Delta state.

This paper goes on to show how Project Gaia's technology transfer-led approach came to replace the more localized model of technology development initially championed by CEHEEN and analyzes the implications of this shift for different categories of energy users. After a review of the practical and methodological considerations which informed the design and execution of the study, the paper chronicles the history of CEHEEN's stove development efforts in the Niger delta before the organization merged with Project Gaia. The paper further elaborates on an observed distinction in the organizational strategy regarding the contexts in which local participation in stove projects is considered desirable, one which signals a movement away from a context-responsive implementation approach toward the privileging of a globally oriented, expert-led approach. The paper concludes with a discussion of the implications of such shifts in favor of global trends for the local responsiveness of the technological solutions that result from North–South collaborative development efforts.

## 19.2 Research Design and Methods

The manner in which case selection proceeded for this study yields significant insight into the rationale underlying the transition of CEHEEN into Project Gaia Nigeria and, by extension, the shifting of the former's focus from the Improved Egaga to the CleanCook project. At the onset of the study, I identified both the Improved Egaga and the CleanCook as projects in which CEHEEN had a stake. However, the Improved Egaga was of primary interest because, having been wholly conceived and implemented by a Southern organization, it appeared to offer lessons in indigenous technology and participatory development, concepts which were at the core of the analysis I intended to undertake.

My inquiry into the Improved Egaga project began with a series of telephone interviews with a senior member of CEHEEN staff over a 7-month period between October 2008 and May 2009. The serial nature of the interviews provided the opportunity to process the information gathered on each occasion and feed my interpretations back into successive conversations. Though I had established over the course of the telephone interviews that the CleanCook project had more or less displaced the Improved Egaga project on CEHEEN's agenda, the organization had given the assurance that it would facilitate access to the communities that had participated in the latter project. Upon arrival in the field, however, I realized that an insistence on studying the Improved Egaga project would be counterproductive to my investigation, as it was apparent that the conditions in the field were not conducive to studying that particular project at the time. Firstly, the structure of the organization had changed significantly following the completion of the Improved Egaga project: CEHEEN, a local NGO, had merged with Project Gaia, an international NGO to become Project Gaia Nigeria, and secondly, the organization, with its new focus on the more recently completed CleanCook pilot project and the proposed commercial follow-up to the project, was unwilling to apportion any substantial amount of time or human resource to the obsolete endeavor they now considered the Improved Egaga project to be.

In response to this rather sudden turn of events, I decided to abandon my initial attempts to stick with the carefully prepared research design and instead build the investigation around the project that Project Gaia Nigeria was interested in at the moment. As such, my attention shifted in the field from the Improved Egaga project to the CleanCook project. The decision to adapt the study to the preference of the project organization constituted a major change in the original research design; however, it was one that had valuable practical and theoretical implications for the research. Practically, negotiation of access to project staff and project communities proceeded at a much faster pace than it did prior to the change in direction. Perhaps more significantly, I came to recognize CEHEEN's abdication of the Improved Egaga project in favor of the CleanCook project as constituting data in itself, thus opening up the interesting line of analysis discussed in this paper.

Fieldwork, consisting of semi-structured interviewing and non-participant observation, was conducted for six weeks between September 2009 and October 2009. A total of six interviews were held at various times with two members of Project Gaia staff who had overseen the CleanCook pilot project in 13 households in Warri, one of the urban centers where the stoves were deployed. Of the 13 project households, it was only possible to gain access to five because the other families involved in the pilot had mostly moved away from the area. In the analysis presented here, the interview data have been supplemented with project information from official documents supplied by Project Gaia Nigeria. The secondary data gathered from these documents provide pertinent insight into the conception and implementation of the discontinued Improved Egaga project as well as the outcomes of the CleanCook pilot in other locations, especially those in rural areas, which could not be accessed during fieldwork due to practical constraints on the organization's time and resources.

## 19.3 The Improved Egaga Stove as Appropriate Technology

In a 2002 CEHEEN report outlining the organization's basis for implementing the Improved Egaga intervention, the gravity of the household energy problem in Nigeria was spelled out as follows: As a consequence of widespread poverty in the country, 80 % of the population, most of whom live in rural communities, relied on fuelwood and other sources of solid biomass to meet their cooking energy needs. In the report, CEHEEN established that this 'unwholesome trend' was further aggravated by the widespread use of inefficient technologies to burn biomass fuels: An estimated 98 % of the biomass used domestically is burnt over open fires or inefficient stoves, with the effect that high levels of indoor air pollution are produced, leading to high mortality and morbidity rates particularly within rural households (Obueh 2002).

To further underscore the magnitude of the problem, the CEHEEN report cited a 'startling' 1992 document published by the Federal Environmental Protection Agency of Nigeria in which about 50 million Nigerians were reported to be directly affected by deforestation due to declining fuelwood supply. Galvanized into action by the apparent seriousness of the problem, CEHEEN launched an 18-month baseline study in 1997 in order to determine cooking energy demand and supply patterns in two communities located in Edo, another state in the Niger delta region. According to Obueh (2002), the results of the baseline study reflected the trends reported by the Federal Environmental Protection Agency and even went on to suggest that the problem actually existed on a far greater scale than was generally assumed: The baseline results showed for instance that in some communities, 'all valuable trees within a 25-km radius had been lost to… an unrestrained exploitation of fuelwood over the years' (Obueh 2002, p. 4). At the time, CEHEEN expressed particular concern over the observation that no coordinated action had yet been taken to address the excessive use of fuelwood by the teeming poor in spite of the threat posed by the trend to the environment. The report concluded at length that the solution to the growing household energy problem in Nigeria was to 'appropriately design cooking devices' (Obueh 2002, p.2) to improve upon the inefficient ones used in poor households across the country.

The preliminary telephone interviews conducted with CEHEEN/Project Gaia Nigeria staff ahead of the main round of fieldwork yielded pertinent data on the contextual realities which informed the decision to implement the Improved Egaga project. In Oghara and Benin, the two baseline study communities, the traditional Egaga stove was identified as the predominant stove used for cooking in the region. Essentially a locally manufactured metal stand used to support a cooking pot over an open fire, the Egaga stove had been in use in the project communities for over 100 years. The bare-bones structure of the traditional Egaga, however, means that much of the fuelwood stacked within its confines is exposed to the open air during cooking. Consequently, when the fuel burns, only a fraction of the heat energy is directed to the cooking pot placed over the fire.

CEHEEN set out to develop 'an appropriate cooking energy technology that potentially uses below 40 % fuel than the traditional technology, and yet attaining more than 50 % efficiency, while also reducing smoke impact by 60 %' (Obueh 2002, p. 3). In approaching this task, CEHEEN worked with two local women's groups, one in each community, to identify the preferences of the mostly female local stove users (Obueh 2001). An improved version of the Egaga stove was developed which was capable of saving up to 40 % of the fuelwood used in the traditional model [Interview Project Gaia Nigeria Staff 1] and conserving up to 60 % of the heat dissipated by the latter (Obueh 2001). Upon testing of three different stove models—the improved Egaga stove, a sawdust-burning stove, and a charcoal-burning stove—in both communities, the improved Egaga stove was identified as the most preferred option (ibid.). The major reason given for its widespread acceptance was the familiarity of the technology leading to ease of adaptation [Interview Project Gaia Nigeria Staff 1]. The ensuing pilot activity saw the improved Egaga stove being disseminated in the two project communities to 5222 households, selected on the basis of people's willingness to be involved and the degree of their susceptibility to the harmful effects of biomass fuel use (ibid.).

The Improved Egaga project was marked by a number of features which are of interest to this study. As can be deduced from the project reports cited above (Obueh 2001, 2002), a degree of local participation was incorporated into the project: Women, who were the end users of the technology, were engaged at some level of decision-making. Further, artisans in the local metalworking industry were involved in the development of the technology and were employed to undertake production of the stoves disseminated in the pilot project (Obueh 2001). In essence, CEHEEN incorporated existing local skills and experience into the development of the project—an approach which, according to Obueh (2002), facilitated the identification of solutions that were appropriate to local requirements and that suited users' sociocultural context. In summary, although it would appear from the accounts given in project reports that the opportunities for local participation could have been harnessed to a greater extent on the Improved Egaga project, it is clear that the project strategy was premised upon a perceived need to start from the existing resources and experiences of poor households in the two pilot communities.

## 19.4 Analysis of a Technological Transition: From the Improved Egaga to the CleanCook

In 2001, the Improved Egaga project received international recognition with an award from the prestigious Ashden Awards for Sustainable Energy based in the UK. That year marked the end of the first phase of the project, following which an ambitious long-term goal was announced by CEHEEN to deliver 17.5 million improved Egaga stoves throughout Nigeria by 2010 (Ashden Awards Web site). In spite of this seemingly good start, however, the Improved Egaga project did not progress beyond the first phase. As indicated in Sect. 19.2, when I got to the field in September 2009 with the intention of investigating the project, I discovered that CEHEEN had put a complete stop to it much earlier and had cut off every contact with the project communities. This made it difficult to unearth up-to-date information about the project. In the course of interviewing ex-CEHEEN-turned Project Gaia Nigeria staff, I experienced difficulty asking questions directly related to the

Improved Egaga project. This was primarily because project staff, who invariably regarded the project as an obsolete subject, were not particularly eager to talk about it. I, however, found it productive to ask questions investigating the reasons behind the shift in organizational focus from the wood-fueled improved Egaga to the alcohol-fueled CleanCook stove.

It is interesting to note that the termination of the Improved Egaga project coincides roughly with the introduction of the CleanCook technology into Nigeria in the early 2000s. CEHEEN initially began its collaboration with Project Gaia as a local partner organization and, as indicated earlier, eventually merged with the latter to pursue the objectives of the CleanCook project in Nigeria. Interview data below show how, following the merger, CEHEEN's original mission gave way to the new one introduced by Project Gaia, an outsider organization which was at the time seeking access points into resource-rich, energy-poor countries for the CleanCook technology. According to ex-CEHEEN staff, the shift in the organization's focus was further legitimated by new developments on the international development scene which prompted a realization of the inadequacy of the improved Egaga technology to tackle the scale of the problem identified for biomass users:

> Just about then, the WHO came out with their report, that over 6 million people—deaths—are recorded every year from the use of biomass stoves. I mean, they said, even the best of biomass stoves still do not, is still not able to alleviate the problem at hand. And that's the problem of alleviating, reducing the incidence of indoor smoke. And, because, stove efficiency in combustion is directly related to how efficient the stove is... Because the target is to cut down smoke completely. The only way you could do that is to use a clean-burning fuel... Of course, if you're going to use a cleaner fuel, you need a technology that will complement the fuel. In other words, you need, you also need clean-burning devices to burn the clean fuels [Interview Project Gaia Nigeria Staff 1].

The latter part of the above statement reflects Project Gaia's prescription regarding what the most appropriate solution to the identified problem would be: a move away from traditional biomass stoves and fuels toward a more modern cooking technology. Local staff of CEHEEN/Project Gaia Nigeria have since 'keyed in' to this expert recommendation and now dismiss the improved Egaga as 'just a biomass stove' [Interview Project Gaia Nigeria Staff 1]. Interview data discussed below show that the improved Egaga stove has evolved in CEHEEN's discourse from a technology that responded appropriately to the energy needs of poor biomass users in Nigeria to one that cannot adequately cater to the needs of the cross section of energy users in the country:

> Biomass stove, we discover, discriminates. Biomass stove can only fit into a particular income group, income strata in the society. So, it discriminates. So, a technology that discriminates as we thought, would not be the best of options in resolving [the] household energy crisis [Interview Project Gaia Nigeria Staff 1].

The point of the above statement is that in light of the varied energy use patterns across the low-, middle-, and high-income groups in Nigeria, the improved Egaga stove 'discriminates' against energy users in the high- and middle-income groups who are higher up the energy ladder than those in the low-income group who are the major users of biomass fuels. Contrasting the improved Egaga technology with the CleanCook technology, another member of Project Gaia Nigeria staff had this to say:

> Egaga was just, it was not representative enough, taking into consideration these criteria that we looked at. [Interview Project Gaia Nigeria Staff 2]

By 'representative,' the above interviewee was referring to the selection criteria used in the pilot phase of the CleanCook project which ensured a spread of participating households across the low-, middle-, and high-income groups in both rural and urban areas of project locations. Some of the responses given by project staff indicate that the move away from the improved Egaga technology was not only a response to the external stimulus provided by international organizations, but also a response to growing pressure from energy users on the ground. The claim is that CEHEEN's transition was also fueled by the enthusiastic response of energy users, including those in poor households, to the CleanCook pilot. Indeed, this discovery of people's desire for a better cooking technology is regarded by project staff as one of the most significant outcomes of the Improved Egaga project. The following interview exchange with Project Gaia Nigeria Staff 1 buttresses this and further reveals a distinction in the significance accorded to local participation in the development and introduction of each technology:

> TS: According to the documentation that I read, there was some element of participation in the Egaga. I'm wondering if the same was present in the CleanCook.
> PGNS: It was more, it was more with the CleanCook.
> TS: Okay, how come?
> PGNS: Yeah, because we felt that more people embraced the CleanCook. They felt it was a better technology, in terms of quality.
> TS: Okay, though the kind of participation I'm talking about, you talked about some women's groups that you worked with, with the Egaga and how they actually contributed to identifying how the technology could be made better.
> PGNS: Oh! Yeah, yeah. The participation was more with the, local participation was more with the Egaga than the CleanCook.
> TS: Why was that?
> PGNS: Yeah, because CleanCook found its market more, I mean it was easier for the middle class, the kerosene users and the LPG gas users to embrace the CleanCook than the Egaga. I think apparently because of the income…

It is interesting to note that the project staff in the exchange above has equated 'embrace' of the CleanCook technology by middle-income households with their participation in the project. This indicates that the focus of 'local participation' for CEHEEN/Project Gaia Nigeria has shifted from poor energy users to those higher up the income pyramid who are better positioned socioeconomically to participate in commercial energy markets. This leads us to another important insight conveyed by the exchange, which is the thinking that the degree to which the local participation is required to facilitate the appropriateness of an external intervention is dependent on the income level of users and their position on the energy ladder. By associating the lower degree of participation by the middle classes in implementation processes with their higher capacity to participate in the CleanCook stove-and-fuel market, the implicit suggestion is that local participation in technology and market development is more desirable as a component of stove projects targeted at low-income households at the bottom of the energy ladder.

The data gathered from CleanCook pilot participants during fieldwork in Warri appear to support the above suggestion. The five women interviewed were all educated beyond secondary level and worked in various professional capacities: In the group, there was a retired accountant, a retired nurse, and two schoolteachers. Consequently, these women have relatively high degrees of socioeconomic empowerment and generally have a wider range of options than do their counterparts in low-income households. The women, who already used kerosene and liquid petroleum gas (LPG) interchangeably depending on cost of the fuels and length of cooking time, welcomed the CleanCook as a cleaner-burning alternative that was 'more economical than using the kero, or the gas' [Interview Warri Female Household Head 2] even though they had been far removed from the process of technology development.

This finding appears to pose a challenge to theories of participation which propound that the development objectives of Northern organizations cannot be fulfilled without the involvement of Southern citizens in the planning and development phases of a project, particularly one that is technological in nature. It should, however, be noted that the acceptance of the CleanCook intervention cited here is situated in a different context than the largely rural, low-income, or subsistence contexts in which the Improved Egaga project, and indeed the majority of improved stove interventions, is implemented in the South. The interview data presented here only account for middle-income urban households which in Nigeria are mostly kerosene users (Bailey et al. 2006); they do not demonstrate the impact of the project on low-income, rural biomass users within the project locations. As explained earlier, I was unable to gain access to rural or urban households classified by the project as being within the low-income range; however, the findings from Project Gaia's 2006 baseline study summarized in Sect. 19.5 give some insight into the impact that the CleanCook intervention can be expected to have on this group of energy users.

## 19.5 Analysis of Differential Impact: The CleanCook and the Energy Ladder

Section 19.4 noted how households in the middle-income group embraced the CleanCook technology despite the non-participatory approach taken by Project Gaia to project development. However, as highlighted earlier, the pilot was designed to span households in various income groups which use fuel sources that correspond to different rungs of the energy ladder, so this response cannot be taken to be representative of the cross section of households involved in the project. This section now proceeds to use the baseline study data—which also cut across low-, middle-, and high-income households in rural and urban locations and thus provide a wider base for analysis—as a benchmark against which to deduce the likely impact of the CleanCook technology on energy poverty alleviation for households within each income group.

Table 19.1 constitutes a graphical representation of the direct relationship between the income pyramid and the energy ladder pinpointed earlier. It can be observed from the table that high-income households are able to spend more on fuel sources higher up the energy ladder than middle-income and low-income households can. The same applies for middle-income households, relative to their low-income counterparts.

Based on the data presented in Table 19.1 alone, it would appear that fuelwood use is comparable between the low-income and middle-income households in the sample. However, as is shown in Table 19.2, the percentage of middle-income (as well as high-income) households that use fuelwood is negligible compared to the percentage of low-income households in that category.

In the course of the pilot, it was determined that a family of four cooking on the CleanCook stove would use 365 liters of ethanol/methanol fuel in a year (Obueh 2004). The cost per liter of ethanol fuel was projected at US$0.38 (Bailey et al. 2006), resulting in an average fuel expenditure of US$138.70 yearly—much higher than the total annual fuel expenditure for fuelwood users across all income groups, as Fig. 19.1 below portrays.

On the other hand, the projected ethanol cost is favorable compared to the current fuel expenditure by kerosene and LPG users, with the exception of kerosene users in the low-income group. Low-income kerosene users are therefore also likely to be exempted from the potential benefits of the CleanCook technology, albeit by a smaller margin than is the case with fuelwood-reliant households across all income groups. Based on the data presented in Tables 19.1 and 19.2, therefore, it appears that the CleanCook will lead to reduced fuel expenditure for households currently on the middle and upper rungs of the energy ladder, but is not likely to alleviate

**Table 19.1** Average annual fuel expenditure by income group (adapted from Bailey et al. 2006)

| Primary household fuel | Average annual fuel expenditure (US$) | | |
|---|---|---|---|
| | Low income | Middle income | High income |
| Fuelwood | 25.09 | 17.90 | 64.43 |
| Kerosene | 95.77 | 317.00 | 681.27 |
| LPG | 294.72 | 431.00 | 1053.60 |

**Table 19.2** Percentage use of fuelwood, kerosene, and LPG by income group (adapted from Bailey et al. 2006)

| Primary household fuel | Percentage use of fuel | | |
|---|---|---|---|
| | Low income (43 homes) | Middle income (67 homes) | High income(25 homes) |
| Fuelwood | 48 % (21/43) | 9 % (6/67) | 4 % (1/25) |
| Kerosene | 48 % (21/43) | 72 % (48/67) | 44 % (11/25) |
| LPG | 2 % (1/43) | 19 % (13/67) | 52 % (13/25) |

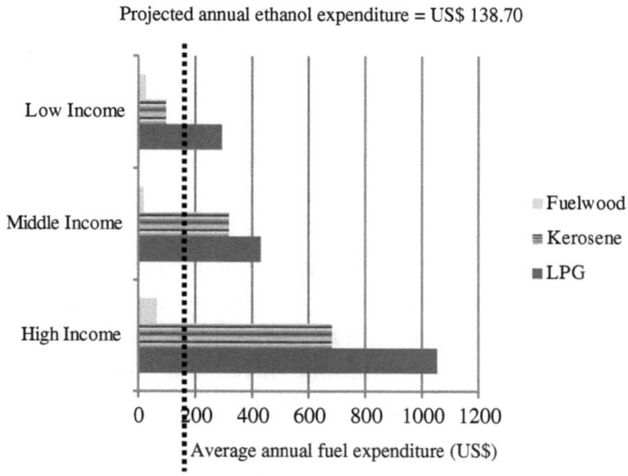


**Fig. 19.1** Fuel type and expenditure by income group (adapted from Bailey et al. 2006)

energy poverty for fuelwood users at the bottom of the energy ladder—those households originally targeted by the Improved Egaga stove and among whom the greatest poverty-alleviating impact could potentially be achieved.

## 19.6 Conclusion

This paper has described the involvement of Project Gaia, a Northern-led NGO, on the Nigerian household energy scene in partnership with CEHEEN, an indigenous organization originally operating based on appropriate technology principles. Importantly, the paper detailed how the advent of Project Gaia's CleanCook stove, framed as a more technologically advanced and more socially inclusive solution than CEHEEN's improved Egaga stove, precipitated a shift from the indigenous technology and participatory principles that CEHEEN had considered so essential to the success of the Improved Egaga project to an expert-led approach characterized by the transfer of a novel stove-and-fuel technology from North to South.

The paper showed how the improved Egaga stove was subsequently relegated by the resulting hybrid organization, Project Gaia Nigeria, on the claim that the technology discriminated against groups higher up the energy ladder than biomass-reliant households. Findings from fieldwork conducted in Warri and analysis of Project Gaia's own findings from a baseline study of energy use patterns across Delta state, however, show that the language of discrimination employed by Project Gaia Nigeria is applicable in the reverse: While women in middle-class households indeed embraced the CleanCook as a welcome addition to the kerosene and LPG fuels that they currently use, those in low-income households who rely overwhelmingly on biomass fuels are less likely to find the technology comparable to, or competitive with, the energy sources they presently use. Indeed, according to the World Health Organization's energy ladder categorization (WHO 2006b), kerosene, LPG, and electricity belong to the middle and upper rungs of the energy ladder along with ethanol and methanol, the principal fuels used in the CleanCook stove. Given the positive relationship established between household energy use and income level in Nigeria, it is perhaps not surprising that middle- and high-income households with access to those middle- and upper-rung fuels are the ones who are positioned to benefit the most from the comparable ethanol-fueled technology.

The above findings show that energy-poor populations in Southern contexts cannot be assumed to be a homogeneous mass. Rather, the energy needs and solutions that would count as essential, and even relevant, for different populations vary across socioeconomic levels—a factor that CEHEEN did not sufficiently consider when redrawing the lines around the problems and technologies it chose to prioritize. Irrespective of their globalizing tendencies, North–South development interactions need to take as their starting point needs and capabilities identified with this context-specific differentiation in view. The outcomes—even if only slightly incremental, as with the Improved Egaga intervention—are likely to be less 'discriminatory' against the poorest who are the primary targets of development efforts.

## References

Ashden Awards. (2010). *Centre for household energy and the environment (CEHEEN), Nigeria: Improved cooking stoves*. http://www.ashdenawards.org/winners/ceheen. Accessed 5 July 2010.

Bailey, S., Samson, R., & Murren, J. (2006). *Baseline data for 150 homes in the communities of Asaba, Abraka and Warri, Delta state, Nigeria*. Resource document. Project Gaia. http://www.projectgaia.com. Accessed 18 February 2010.

Leach, G. (1992). The energy transition. *Energy Policy, 20*(2), 116–123.

Obueh, J. (2001). Using a household energy technology to promote small-scale enterprises in rural communities in Nigeria—The Egaga stove experience. *Boiling Point, 47*, 28–29.

Obueh, J. (2002). *Promotion of energy-efficient cooking stoves for the rural poor in Nigeria: Special report prepared for the World Bank Global Development Network Awards 2002*. Nigeria: Centre for Household Energy and the Environment.

Obueh, J. (2004). *Project Gaia research studies Delta state, Nigeria: Mini-pilot study—A test run in anticipation of a full study*. Resource document. Project Gaia. http://projectgaia.com/files/MiniCleanCookPilotStudyDeltaState.pdf. Accessed 10 April 2010.

Pachauri, S., & Spreng, D. (2003). *Energy use and energy access in relation to poverty*. Resource document. Centre for Energy Policy and Economics. http://www.cepe.ethz.ch/publications/workingPapers/CEPE_WP25.pdf. Accessed 13 January 2014.

Stokes, H., & Ebbeson, B. (2005). Project Gaia: Commercializing a new stove and new fuel in Africa. *Boiling Point, 50*, 31–33.

WHO. (2006a). *Global health atlas: World health statistics*. Geneva: World Health Organization.

WHO. (2006b). *Fuel for life: Household energy and health*. Geneva: World Health Organization.

World Bank. (2013). Removing the smoke from making tortillas in Central America. http://www.worldbank.org/en/news/feature/2013/09/13/smoke-from-tortilla-making-in-central-america. Accessed 9 January 2013.

# Part VI
# Targeted Training and Capacity Building Energy Programs

# Chapter 20
# Supporting the Development and Deployment of Sustainable Energy Technologies Through Targeted Scientific Training

**Jennifer M. MacLeod and Federico Rosei**

**Abstract** Sustainable energy technologies rely heavily on advanced materials and modern engineering controls. These promising new technologies cannot be reliably deployed without ensuring that there is a sufficient "capacity," i.e., trained technical personnel with the expertise to implement, monitor, and maintain the energy infrastructure. This same capacity is critical to the local development of new technologies, especially those that respond directly to regional priorities, strengths, and needs. One way to build capacity is through targeted programs that integrate the training and development of locals at an advanced technical level. In practical terms, these programs usually produce a small number of highly educated individuals with skills in science and engineering. The goal of Part VI of this book is to highlight contributing factors in successfully operating capacity building programs.

## 20.1 Capacity Building to Foster the Dissemination and Development of Sustainable Energy Technology

The widespread dissemination of sustainable energy technologies is a global imperative. These technologies represent an opportunity to reduce greenhouse gas emissions while at the same time increasing access to energy in regions where traditional energy infrastructure is expensive, unreliable, or has not been established. Sustainable energy technologies are often well suited to small-scale implementations, obviating the need for centralized power generation and transmission, and are therefore excellent candidates for creating energy access where it is

J.M. MacLeod (✉) · F. Rosei
UNESCO Chair in Materials and Technologies for Energy Conversion, Saving and Storage, Institut Nationale de la Recherche Scientifique (INRS), Varennes, Canada
e-mail: macleod@emt.inrs.ca

F. Rosei
e-mail: rosei@emt.inrs.ca



S. Hostettler et al. (eds.), *Sustainable Access to Energy in the Global South*,
DOI 10.1007/978-3-319-20209-9_20

most needed. The importance of access to energy is profound: energy enables all aspects of development, and sustainable energy is a defined priority in the United Nations Development Programme (Takada and Fracchia 2007).

Knowledge and capacity gaps are among the most important challenges that hinder the dissemination of sustainable energy technologies in less-developed nations (Rosei et al. 2008; Kruckenberg 2014). The successful implementation of novel energy technologies requires a critical mass of trained individuals who can support and maintain the required infrastructure (Terrapon-Pfaff et al. 2012). In a richly supportive environment, i.e., one with sufficient technical capacity, technologies can be adapted and improved over time to reflect regional context and the needs of the end users. This type of environment can also serve as an incubator for new technologies and infrastructure (Bell and Pavitt 1997).

The focus of Part VI of this book is on the capacity building and technology and knowledge transfer aspects of small-scale initiatives in energy technology in less-developed nations. In general, contemporary technology deployment programs recognize the importance of capacity building for their long-term viability (Chaurey et al. 2012). However, the planning and critical evaluation of capacity building initiatives is a notoriously difficult undertaking, and the relative novelty of such programs within the context of sustainable energy creates a particular barrier to their evaluation and evidence-based design, since long-range outcomes are unknown (Wing 2004). Additionally, small programs may be sufficiently different from one another so as to be essentially unique, making it difficult to apply a standard rubric for evaluating outcomes. We emphasize instead the benefit of learning from case studies, the preliminary identification of best practices based on short-term outcomes, and the examination of individual program successes and challenges to uncover their universal applicability.

For instance, in Chap. 21, entitled *Building Local Capacities to Monitor Methane Extraction in Lake Kivu*, Natacha Pasche et al. discuss the training of local Rwandese technicians for the Lake Kivu Monitoring Program, which is designed to ensure that the ecosystem and the stability of the lake are preserved during methane extraction operations. In particular, the authors describe successes relating to the technicians' extended education through participation in courses, conferences and workshops, and team-building through a thoughtful managerial approach. The technicians also benefitted from extended contact with and instruction from an expert in the field. These types of extended interactions and pedagogic opportunities provide a framework of "strong ties" for capacity building (Granovetter 1973). The use of strong ties in capacity building programs is often identified as or assumed to be desirable, although operational programs may fall short of successfully implementing this paradigm (Kruckenberg 2014).

In Chap. 22 entitled *Bali, Indonesia: Combating Climate Change and Poverty—Recycling Used Cooking Oil by Transforming it into Biodiesel*, Thorsten Reckerzügl alludes to an important challenge to supporting technical training of personnel in alternative energy programs: strong management structures must be in place for these programs to flourish, and these management structures also require local capacity. The strong ties facilitated at the Lake Kivu Monitoring Program result are correlated with strong program management and careful design of the technical training and support elements.

These case studies provide us with a window into the operational aspects of capacity building initiatives in sustainable energy technologies and contribute to what will surely be a growing literature describing such programs. For those of us working on the practical implementation of capacity building in sustainable energy, these case studies are a valuable resource. As part of the UNESCO Chair in Materials and Technologies for Energy Conversion, Saving and Storage (MATECSS), we have undertaken our own training programs, designed to increase research capacity in less-developed nations primarily through visiting professorships and training of students at the doctoral level. These types of young, small programs are ideally positioned to benefit from the lessons of operational programs, and to emphasize and support design considerations that are associated with positive outcomes, such as those presented in Part VI of this book.

## References

Bell, M., & Pavitt, K. (1997). Technological accumulation and industrial growth: Contrasts between developed and developing countries. In D. Archibugi & J. Michie (Eds.), *Technology, globalisation and economic performance* (pp. 83–137). Cambridge: Cambridge University Press.

Chaurey, A., Krithikaa, P. R., Palita, D., Rakesha, S., & Sovacoolb, B. K. (2012). New partnerships and business models for facilitating energy access. *Energy Policy, 47*(s1), 48–55.

Granovetter, M. S. (1973). The strength of weak ties. *American Journal of Sociology, 78*(6), 1360–1380.

Kruckenberg, L. A. (2014). Empowering partnerships? Renewable energy partnerships in development cooperation. *Paper presented at Tech4Dev 2014*. Lausanne: EPFL. June 2014.

Rosei, F., Vayssieres, L., & Mensah, P. (2008). Materials science in the developing world: Challenges and perspectives for Africa. *Advanced Materials, 20*, 4627–4640. doi:10.1002/adma.200802222.

Takada, M., & Fracchia, S. (lead authors) (2007). *A review of energy in national MDG reports*. New York: United Nations Development Programme. http://www.undp.org/content/dam/aplaws/publication/en/publications/environment-energy/www-ee-library/sustainable-energy/a-review-of-energy-in-national-mdg-reports/A%20Review%20of%20Energy%20in%20National%20MDG%20Reports_2007.pdf. Accessed March 22, 2015.

Terrapon-Pfaff, J., Dienst, C., König, J., & Ortiz, W. (2012). A cross-sectional review: Impacts and sustainability of small-scale renewable energy projects in developing countries. *Renewable and Sustainable Energy Reviews, 40*, 1–10.

Wing, K. T. (2004). Assessing the effectiveness of capacity-building initiatives: Seven issues for the field. *Nonprofit and Voluntary Sector Quarterly, 33*(1), 153–160. doi:10.1177/0899764003261518.

# Chapter 21
# Building Local Capacities to Monitor Methane Extraction in Lake Kivu

**Natacha Pasche, Janvière Tuyisenge, Ange Mugisha, Edouard Rugema, Alice Muzana, Aline Uwasempabuka and Augusta Umutoni**

**Abstract** Lake Kivu is a large tropical lake situated in Africa between Rwanda and Congo. This unique lake contains an enormous quantity of dissolved methane and carbon dioxide. Their sudden release could have catastrophic consequences for the two million inhabitants of its riparian areas. To avoid such a risk, the Rwandan and Congolese governments have decided to extract the methane. Methane extraction will produce as much as 500–800 MW of energy over the next 50 years. Then, the extraction will be reduced to match the natural rate of methane replenishment, transforming it into a renewable source of energy. This innovative win-win technology is essential for the development of Rwanda that is currently producing 100 MW of energy and where only 16 % of the population has access to electricity. In 2008, the first pilot plant started to extract and burn methane, generating ~1.3 MW. To accompany this technological success, the Government of Rwanda created the Lake Kivu Monitoring Program (LKMP) to ensure its safe and environmental friendly exploitation. This team is monitoring the main risks associated with methane extraction, namely the alteration of the lake stability and the potential deterioration of the lake ecosystem. LKMP was financially supported by North-South cooperation through two successive capacity building projects. For four years, an international expert trained five Rwandan technicians to become autonomous and set up a laboratory for nutrients, gases, and biological analyses. This local team learned to detect potential impacts on the lake ecology and on the safety of the population. Although knowledge transfer and practical experiences were successful, the outcome remains fragile and the local team still needs to be reinforced. At the regulatory level, the institutional framework remained weak, and the laws for extraction still need to be enforced. In the future, these challenges should be addressed to ensure adequate monitoring with strong local capacities.

N. Pasche (✉)
Limnology Center, Ecole Polytechnique Fédérale de Lausanne, Lausanne, Switzerland
e-mail: natacha.tofield-pasche@epfl.ch

N. Pasche · J. Tuyisenge · A. Mugisha · E. Rugema · A. Muzana · A. Uwasempabuka
A. Umutoni
Lake Kivu Monitoring Program, Energy and Water Sanitation Authority, Kigali, Rwanda



S. Hostettler et al. (eds.), *Sustainable Access to Energy in the Global South*,
DOI 10.1007/978-3-319-20209-9_21

## 21.1 Introduction and Purpose

For two decades, Rwanda has experienced an energy crisis due to a lack of investment in the energy sector. High prices, low power production, and limited accessibility to electricity have restricted the economic development of the country. In 2007, only 4.3 % of the 9.1 million inhabitants had access to electricity with a total installed capacity of 75 MW. The electricity consumption per capita (30 kW h/year/inhabitant) is among the lowest in the world (Safari 2010). Between 2008 and 2011, the Government of Rwanda has expanded access to electricity by an impressive 160 %, but still only 16 % of Rwanda's households are connected to the electricity grid. The power supply in 2012 was generated on 100 MW of installed capacity. In 2012, the government set a national target to increase access to electricity by 70 % and electricity production to 1160 MW by 2017. This new capacity should include the production of 300 MW of energy from methane from Lake Kivu (Ministry of Infrastructure 2011). The benefits of methane extraction are twofold, as it will simultaneously decrease the risk of eruption in the lake and provide electricity.

Lake Kivu, a large tropical lake, is situated in the active Great African Rift Valley between the Republic of Rwanda and the Democratic Republic of the Congo (DRC). Largely influenced by two active volcanoes, Nyiragongo and Nyamulagira, this unique lake contains an enormous quantity of dissolved methane [$\sim$60 million cubic meters ($m^3$)] and carbon dioxide ($CO_2$ $\sim$ 300 million $m^3$). Worldwide, only two other lakes contain high concentrations of $CO_2$ in their deep water. In 1984 and 1986, landslides triggered spontaneous gas eruptions from Lake Monoon and Lake Nyos, which asphyxiated 34 and 1740 inhabitants of the surrounding areas, respectively (Kling et al. 1987). These unfortunate eruptions occurred because the $CO_2$ concentrations in the lake were near 100 % saturation. Contrary to these two "killer lakes," Lake Kivu currently has a maximum gas saturation of 55 %, and gases are efficiently trapped below 260 m. To trigger an eruption, water from 320 m would have to rise by 150 m, before gas bubbles start forming spontaneously (Schmid et al. 2004). In 2005, the lake was considered to be safe, but natural methane accumulation could become dangerous in the next 100 years (Schmid et al. 2005). Therefore, methane extraction is essential to reduce this risk in the long term.

Depending on the efficiency of the process, methane extraction could produce as much as 500–800 MW of energy over the next 50 years. Then, the extraction should be reduced to the annual natural replenishment rate of methane in the lake, transforming it into a veritable renewable source of energy. Since October 2008, methane has been extracted by the Kibuye Power 1 (KP1) pilot plant, which generates $\sim$1.3 MW of energy and has a 3.6 MW installed capacity. This government-owned company is in the process of being acquired by the private company Israel Africa for a total concession of 50 MW. Another pilot station of 25 MW, which was constructed by KivuWatt, should start extracting by 2015 near the town of Kibuye. This American company has already signed a concession for a total of 100 MW with the Government of Rwanda. In addition, the DRC and Rwanda have also initiated a project for a common exploitation of 200 MW.

To ensure a safe and environmental friendly exploitation, the LKMP was set up under the Ministry of Infrastructure in 2008 and was then transferred to the Energy and Water Sanitation Authority (EWSA) in 2011. The team is monitoring the main risks associated with methane extraction: an alteration of the lake stability and a deterioration of the lake ecosystem due to an increase in nutrient inputs. To minimize these risks, international experts elaborated rules and guidelines for extraction in the document entitled "Management Prescriptions for the Development of the Lake Kivu Gas Resource" (Boyle et al. 2010), which is based on the modeling of different extraction scenarios (Wüest et al. 2012). The creation of a bilateral regulation authority has been initiated between Rwanda and the DRC, to enforce that these prescriptions are followed and to jointly manage the methane resource.

In this paper, we will briefly explain the principles of methane extraction and then focus on the methodology used within the LKMP. We will describe some results from the near-plant and lake-wide monitoring, focusing on the capacity building of the local technician team. The conclusion will provide an outlook on the way forward.

## 21.2 Design and Methods

Since October 2008, the KP1 pilot station has extracted methane by pumping gas-rich water from a depth of 320 m (Fig. 21.1). After a first separation near the surface, the raw gas, containing approximately 30 % of methane, is transferred to a scrubber, where it is washed. As $CO_2$ is more soluble than methane, part of the $CO_2$ is re-dissolved in washing water that is taken from a depth of 10 m. After the washing process, the enriched gas has a concentration of around 50 % methane and 50 % $CO_2$. This "sweet gas" is then dried and sent through a pipeline onshore, where it is burned in generators to produce electricity. The degassed deep water and the washing water are then combined and re-injected into the lake at a depth of 90 m. Most of the potential impacts on the lake originate from this re-injection that is closely surveyed by the LKMP.

The objectives of the LKMP can be divided into three different levels. First, on-plant inspections ensure that methane-extracting companies are respecting the elaborated rules for a safe extraction. On a second level, the LKMP has measured vertical physicochemical profiles in the lake around the platform to check the impacts of the re-injected water. Through this near-plant monitoring, the signal of the re-injected water was detected using an underwater probe (Sea & Sun Technology, CTD60M probe) that measures depth, pH, temperature, conductivity, and turbidity. On a third level, a lake-wide, long-term program monitors the lake density stratification, ensures the ecosystem integrity, and follows the gas resource. To complement this applied monitoring, the LKMP encourages further research by regional and international scientists to improve the current knowledge of Lake Kivu.

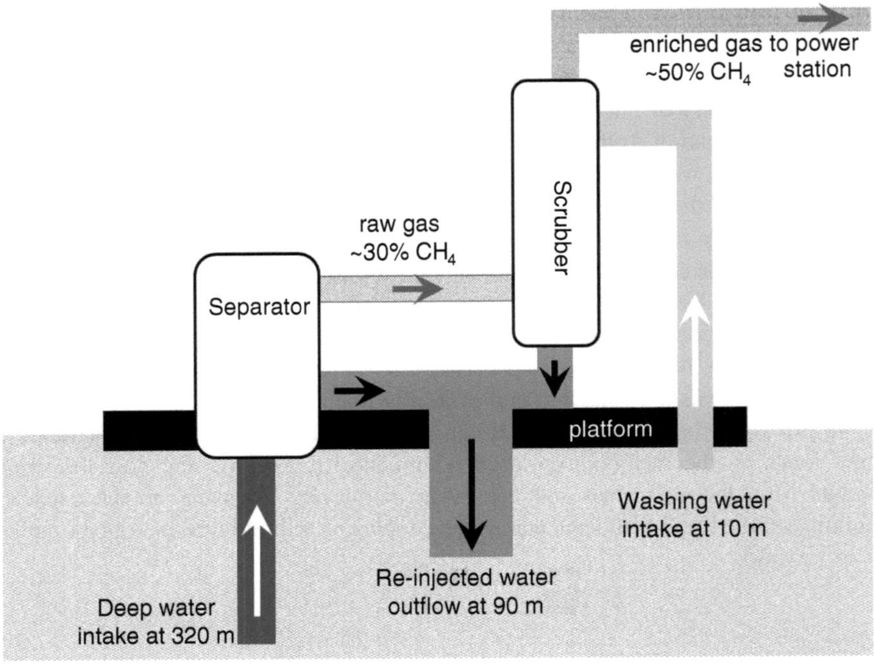


**Fig. 21.1** Schematic design of methane extraction from the KP1 pilot plant. Water pipes are represented in *plain gray,* while gas pipes are represented with *gray dots*. Different *arrows* represent the flow direction

The Rwandan Government aimed to build a strong, local team that can efficiently monitor the impact of methane extraction. The capacity building objectives are threefold: to improve the skills of their employees, to reinforce the organization within the team, and to build up a strong institutional framework in collaboration with external institutions. To accomplish these objectives, an international expert in limnology was employed for four years. This position was financed through two capacity building projects, namely the Investment Climate Facility for Africa and the Public Sector Capacity Building Secretariat. In addition, the Netherlands Embassy in Rwanda provided US$1 million to set up a laboratory.

The expert in limnology trained five Rwandan technicians to detect potential impacts from methane extraction in Lake Kivu. These technicians acquired necessary scientific skills mainly via hands-on training, as well as by collecting and interpreting data in reports. The local team is now experienced in fieldwork such as the collection of physicochemical profiles, deepwater and gas samples, zooplankton, and sedimenting material. In the laboratory, concentrations of nutrients (ammonium, phosphate, and silica) and oxygen, as well as alkalinity, are analyzed by standard chemical methods, and the concentrations of methane and $CO_2$ are determined by gas chromatography. The local technicians have also regularly prepared samples for phytoplankton pigment analyses and preserved zooplankton samples.

The collected data were also interpreted and reported in monthly reports, as well as a more in-depth analysis in an annual report. Their knowledge on Lake Kivu was also reinforced by participating in workshops, international conferences, and limnology courses provided by the expert. Specific training was also provided for phytoplankton analyses at the University of Namur in Belgium, for nutrients analyses and gas sampling at the Eawag research institute in Switzerland, and for hydro-acoustic surveys at Centre Alpin de Recherche sur les Réseaux Trophiques des Ecosystèmes Liminiques (CARRTEL) at Thonon-les-Bains in France. Therefore, the team has developed essential skills and gained strong practical experience to monitor methane extraction.

In addition to lake monitoring and scientific training, particular care was taken to establish an efficient management framework. Clear protocols were developed, and data quality was controlled independently at every work stage. This setup, as well as the defined procedures, should ensure continuous monitoring activities during future personnel changes. The internal organization proved efficient, but the lack of regulations at the institutional level prevented the LKMP from enforcing the rules for safe methane extraction. This challenge needs to be addressed in the near future.

## 21.3 Results

In this section, we will first discuss the results of the near-plant monitoring that focused on the impacts of the re-injected water on the lake around the extracting plant. In the second part, we will describe the results of the lake-wide monitoring program that aims at characterizing the lake's ecosystem, as well as its stability and gas resources.

### *21.3.1 Near-Plant Monitoring*

Seventy campaigns around the KP1 pilot plant were conducted to detect and localize the plume of re-injected water. For each campaign, the Rwandan technicians took deep profiles of temperature, conductivity, pH, and turbidity. Approximately 19 profiles per campaign were collected in one to three circles, varying in radius from 0 to 1200 m, around the platform. The signal of re-injected water was most visible through peaks in the temperature and pH profiles, while conductivity was only affected close to the station (Fig. 21.2). The signal usually decreased with distance from the plant and was observed up to 1.2 km away (Pasche et al. 2011). The plume was often observed toward the shores in the east and rarely toward the open lake in the west. However, the main direction of the plume seemed to vary with time toward the north or the south, depending on wind conditions. In conclusion, the re-injected water affected a surface area of 4.5 $km^2$, which is not even 0.2 % of the lake surface (2370 $km^2$). In addition to the horizontal

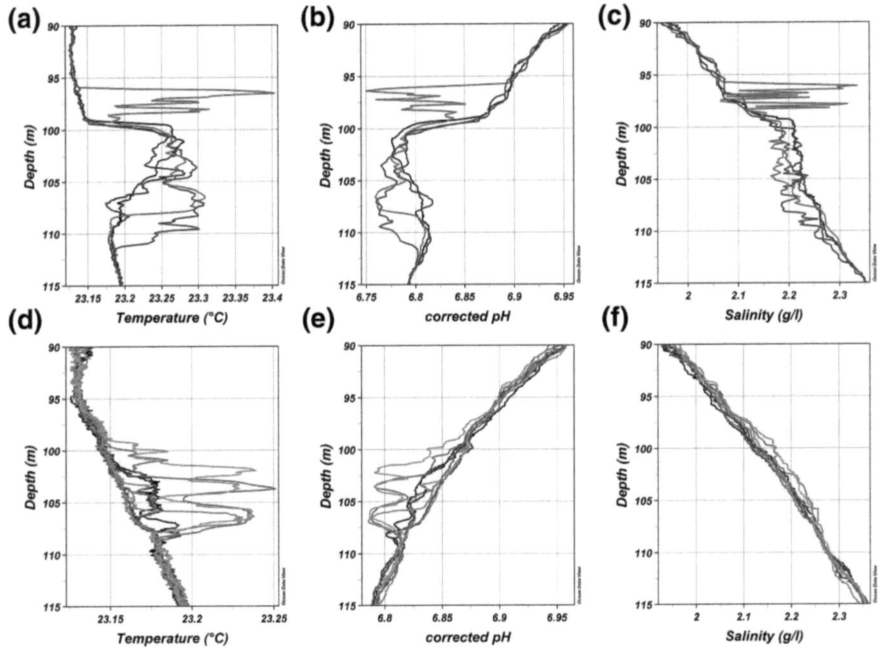


**Fig. 21.2** Temperature (°C), corrected pH, and salinity ($gl^{-1}$) from 90 to 115 m around two circles at 0 m (**a–c**) and 300 m (**d–f**) from the KP1 pilot plant measured on November 28, 2012

extension, it is important to understand the depth at which the re-injected water stabilizes in the water column. Indeed, the main concern about the re-injected water was a local destabilization of the water column. Our hypothesis was that as the density of the re-injected water equals the density of the lake water at 250 m, the re-injected water should sink before stabilizing at 250 m. In reality, we observed that the re-injected water stabilized at around 105 m. This reduced sinking is caused by a local dilution with lake water that we estimated to be a factor of 10. Therefore, this local dilution is crucial for the stability of the lake (Pasche et al. 2013a). We will now describe the magnitude of the nutrient release from the re-injected water.

The nutrient-rich, re-injected water increased nutrients by 1–3 % compared with natural nutrients inputs. Indeed, the nutrient concentrations in Lake Kivu are increasing stepwise with depth (Pasche 2009). Phosphate at 320 m (5.46 mg P–PO4/l) is more than eight times more concentrated than at the reinjection depth (0.66 P–PO4/l). The nutrient cycles of Lake Kivu (Pasche et al. 2012) indicated that the majority of nutrients are released in the mixed layer through a slow upwelling. This internal loading accounts for 90 % of phosphate, 75 % of dissolved nitrogen, and 55 % of silica sources. Thus, external inputs via rivers and atmospheric inputs proved to be of lesser importance (Muvundja et al. 2009). We measured the nutrient concentrations in the re-injected water and compared them to those of the natural inputs. We estimated total inputs by multiplying the discharge by the nutrient

**Table 21.1** Nutrients inputs for the KP1 pilot plant compared to natural inputs in Lake Kivu (Pasche et al. 2013a)

| | $N–NH_4$ (t) | $P–PO_4$ (t) | $Si–SiO_2$ (t) |
|---|---|---|---|
| Normalized load of KP1 per year | 486 | 50 | 370 |
| External nutrients per year | 2600 | 230 | 24,600 |
| Upward flux of nutrients per year | 18,500 | 1800 | 29,600 |
| % external nutrients | 18.7 | 21.7 | 1.5 |
| % upward fluxes | 2.6 | 2.8 | 1.2 |

concentrations. These results (Table 21.1) showed that the artificial inputs account for only 2.8 % of phosphate, 2.6 % of nitrogen, and only 1.2 % of silica. These increases are probably within the annual variability observed due to the changing intensity of the annual deepwater mixing. However, these percentages remain low only because of the small discharge of the pilot plant. It is important to note that such an increase in nutrients will not be tolerated for industrial plants. In fact, the management prescriptions proposed to have two re-injection systems. The washing water should be re-injected at 70 m, while the degassed deep water should be re-injected at 270 m (Boyle et al. 2010). With this system, the annual inputs of nutrients will not be changed, even with a large extraction over the next 50 years (Wüest et al. 2012).

### 21.3.2 Lake-Wide Monitoring by the Lake Kivu Monitoring Laboratory

To reinforce the LKMP, a laboratory was set up for the analyses of gases, nutrients, and plankton near Lake Kivu. The laboratory equipment and a research boat were bought for approximately US$1 million. This funding, provided by the Dutch Embassy in Rwanda, was managed by the World Bank under the Sustainable Energy Development Project. The main instruments installed in 2012 included a Perkin Elmer gas chromatograph with a headspace sampler to measure methane and carbon dioxide, a spectrophotometer to analyze nutrients, a water distillator, a titrator to measure oxygen and alkalinity, two microscopes, two meteorological stations, a fume hood and dishwasher, as well as glassware, chemicals, and small laboratory materials. Therefore, the LKMP is fully equipped to efficiently analyze the necessary parameters to monitor methane extraction.

The maintenance of this sophisticated equipment is challenging. Indeed, expertise and spare parts are not available locally and must be provided by external, developed countries. Local procurement and administrative issues need to be addressed to ensure a functional laboratory. For its economic sustainability, the LKMP officially provides services to other local institutions or scientists for defined prices. The generated funding should allow the LKMP to replace spare parts, as well as to ensure the maintenance of the instruments over the years.

With this new laboratory, the LKMP wrote the first dedicated report for lake-wide monitoring in 2013 (Pasche et al. 2013b), which included meteorological datasets, a comparison of lake stratification, deep profiles of nutrients, as well as the gas concentrations in Lake Kivu. The second part, which focused on plankton and fish-stock assessments, was the first scientific outcome from the project with Belgian and French researchers. This biological baseline study will provide, in combination with previous studies (Sarmento et al. 2006; Isumbisho et al. 2006; Guillard et al. 2012), a robust representation of the lake ecosystem. This baseline will be important in the future to detect any potential deterioration of the ecosystem due to methane extraction.

## 21.4 Conclusion

Methane extraction in Lake Kivu is an essential technology for Rwanda as it will both supply electricity and reduce the danger of a limnic eruption. However, this new and unique extraction technology still needs to be optimized to improve its efficiency. In parallel, the potential impacts on the lake and on the safety of the local population are carefully monitored by local technicians in the LKMP.

The capacity building of the LKMP was strongly supported by the Government of Rwanda and by different sponsors from developed countries. During two capacity building projects, knowledge transfer was mainly performed by an expert in limnology via specific training courses. The presence of an international expert on-site was undeniably beneficial for the development of the LKMP. This expert had a strong knowledge of Lake Kivu, having already investigated nutrient cycles and methane production in the lake (Pasche 2009) during doctoral studies. With such a background, knowledge was efficiently transferred to a team of technicians that grew from two to five. Necessary knowledge in fieldwork, laboratory analyses, and data interpretation was acquired through hands-on experiences. With regular practice, the team became autonomous and skilled in performing the different monitoring tasks. They are able to detect strong impacts on the lake ecology and on population safety. Specific courses abroad also helped to reinforce the technicians' skills.

Building local capacities for the LKMP also generates challenges that will need to be addressed to ensure long-term monitoring. The main challenges are to maintain the specialized laboratory equipment, to ensure the long-term quality and management of collected data, and to resolve administrative issues that lead to long delays and, consequently, missing data. Directly confronted by local realities, the expert helps to resolve administrative problems and to reduce long administrative processes to ensure continuous data collection. However, constant reforms within governmental institutions and rapid changes in key decision-makers caused discontinuities in management, which resulted in repetitive delays and frustrations. To deal with day-to-day running expenses, a more stable and autonomous administration with more dynamic procurement processes should be put into place.

On the institutional level, the current framework remains weak, and the rules for methane extraction are still not enforced by both governments. In 2014, the management prescriptions for safe extraction will be reviewed by external experts thanks to funding from the European Union. Implementing such regulations is necessary to empower both governments and the LKMP when dealing with methane extraction companies.

Even though continuous knowledge transfer and practical experience were successful, the local team still needs to be reinforced after four years of training. The strengthening of the LKMP was requested by the Government of Rwanda because the resources of Lake Kivu need to be preserved as methane production increases. Therefore, the Dutch Embassy has agreed to finance a three-year program to reinforce the LKMP. With a total of US$9.2 million, the team will grow to 15 persons, including an additional technician, several administrative assistants, as well as two new experts in the field of limnology and on-plant inspections. The LKMP should become more independent from the EWSA, and it will hopefully become integrated into a future bilateral authority. Consequently, the LKMP could provide local capacities to ensure the long-term local monitoring of methane extraction.

## References


Boyle, J., Hirslund, F., Morkel, P., Schmid, M., Tietze, K., & Wüest, A. (2010). *Management prescriptions for the development of Lake Kivu gas resources, prepared for Ministry of Infrastructure Republic of Rwanda and ministry of hydrocarbons democratic Republic of the Congo*. http://www.lake-kivu.org/sites/default/files/Kivu%20Management%20Prescriptions%20-%20Final%20Version%207%20jan%202010_0_0.pdf. Accessed 7 December 2013.

Guillard, J., Darchambeau, F., Mulungula, P. M., & Descy, J.-P. (2012). Is the fishery of the introduced Tanganyika sardine (Limnothrissa miodon) in Lake Kivu (East Africa) sustainable? *Journal of Great Lakes Research, 38*(3), 524–533.

Isumbisho, M., Sarmento, H., Kaningini, B., Micha, J.-C., & Descy, J.-P. (2006). Zooplankton of Lake Kivu, East Africa, half a century after the Tanganyika sardine introduction. *Journal of Plankton Research, 28*(11), 971–989.

Kling, G. W., Clark, M. A., Compton, H. R., Devine, J. D., Evans, W. C., Humphrey, A. M., et al. (1987). The 1986 Lake Nyos gas disaster in Cameroon, West Africa. *Science, 236*(4798), 169–175.

Ministry of Infrastructure, Republic of Rwanda. (2011). *National energy policy and strategy*. https://energypedia.info/images/7/77/ENERGY_POLICY_and_STRATEGY.pdf. Accessed 7 December 2013.

Muvundja, F., Pasche, N., Bugenyi, F. W. B., Isumbisho, M., Müller, B., Namugize, J.-N., et al. (2009). Balancing nutrient inputs to Lake Kivu. *Journal of Great Lakes Research, 35*(3), 406–415.

Pasche, N. (2009). *Nutrient cycling and methane production in Lake Kivu*, Dissertation, 127 pp., ETH No.18606, Zürich.

Pasche, N., Mugisha, A., Rugema, E., & Umutoni, A. (2011). *Monitoring the effects of methane extraction in Lake Kivu: May 2010 to April 2011*. 30 pp. Ministry of Infrastructure, Kigali.

Pasche, N., Muvundja, F., Schmid, M., Wüest, A., & Müller, B. (2012). Nutrient cycling in Lake Kivu. In J.-P. Descy, et al. (Eds.), *Lake Kivu: Limnology and biogeochemistry of a tropical great lake* (pp. 31–35). Dordrecht/Heidelberg/New York/London: Springer.

Pasche, N., Mugisha, A., Uwasempabuka, A., & Tuyisenge, J. (2013a). *Monitoring the effects of methane extraction in Lake Kivu: Near-plant monitoring and on-plant inspections, May 2012 to April 2013*. 38 pp. EWSA, Kigali.

Pasche, N., Mugisha, A., Muzana, A., Rugema, R., Uwasempabuka A., & Tuyisenge, J. (2013b). *Lake Kivu monitoring program: Lake-wide monitoring May 2012 to April 2013*, 63 pp. EWSA, Kigali.

Safari, B. (2010). A review of energy in Rwanda. *Renewable and Sustainable Energy Reviews, 14* (1), 524–529.

Sarmento, H., Isumbisho, M., & Descy, J.-P. (2006). Phytoplankton ecology in Lake Kivu (Eastern Africa). *Journal of Plankton Research, 28*(9), 815–829.

Schmid, M., Tietze, K., Halbwachs, M., Lorke, A., McGinnis, D., & Wüest, A. (2004). How hazardous is the gas accumulation in Lake Kivu? Arguments for a risk assessment in light of the Nyiragongo volcano eruption of 2002. *Acta Vulcanologica, 14/15*(1–2), 115–121.

Schmid, M., Halbwachs, M., Wehrli, B., & Wüest, A. (2005). Weak mixing in Lake Kivu: New insights indicate increasing risk of uncontrolled gas eruption. *Geochemistry, Geophysics, and Geosystems, 6*(7). doi:10.1029/2004GC000892 Q07009.

Wüest, A., Jarc, L., Bürgmann, H., Pasche, N., & Schmid, M. (2012). Methane formation and future extraction in Lake Kivu. In J.-P. Descy, et al. (Eds.), *Lake Kivu: Limnology and biogeochemistry of a tropical great lake* (pp. 165–180). Dordrecht/Heidelberg/New York/London: Springer.

# Chapter 22
# Bali, Indonesia: Combating Climate Change and Poverty—Recycling Used Cooking Oil by Transforming It into Biodiesel

**Thorsten Reckerzügl**

**Abstract** Recycling residual materials in a sustainable way is a challenge many developing and emerging countries are faced with. Against this backdrop, Caritas Switzerland (CaCH) has developed a used cooking oil (UCO) recycling project in Bali, Indonesia. The objective of the project is to recycle UCO from hotel kitchens and restaurants and transform it into biodiesel. The biodiesel will in turn be used as a substitute for fossil fuel, thereby reducing greenhouse gas emissions. As part of the project, a local social enterprise has been founded that manages the recycling of UCO. Besides climate change mitigation, the project also contributes to the protection of the local environment and the reduction of human health risks. Furthermore, it provides income and training to socially disadvantaged people. In this chapter, the project's approach is explained and the results achieved as well as the challenges faced are examined. Before drawing final conclusions, the existing constraints are analyzed from the perspective of public service provision.

## 22.1 Introduction and Purpose

The objective of the project described in this chapter is to contribute to climate change mitigation and environmental and health protection by looking at recycling the used cooking oil (UCO) into biodiesel. The biodiesel will in turn substitute fossil fuel, thereby reducing greenhouse gas emissions. The project was set up between September 2010 and July 2013 by Caritas Switzerland (CaCH) and its partners, the Municipality of Denpasar, myclimate, and KUONI Travel Holding Ltd. As part of the project, the local social enterprise, Yayasan Lengis Hijau (YLH), has been founded to oversee the collection and processing of UCO from hotels and restaurants.

---

T. Reckerzügl (✉)
Caritas Switzerland, Lucerne, Switzerland
e-mail: treckerzuegl@caritas.ch

 
S. Hostettler et al. (eds.), *Sustainable Access to Energy in the Global South*,
DOI 10.1007/978-3-319-20209-9_22

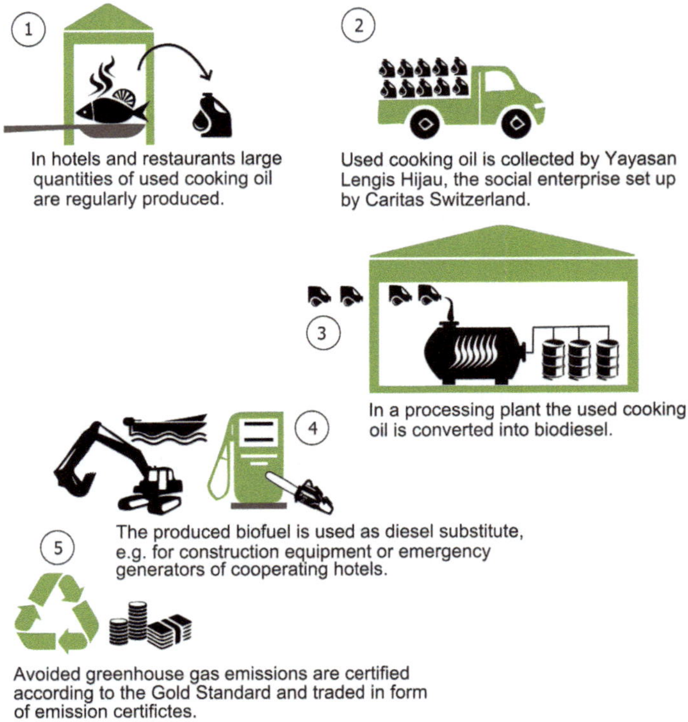


**Fig. 22.1** Process scheme of UCO recycling

Figure 22.1 illustrates the different steps of the process. They are described in more detail as follows:

- Hotels and restaurants collect and store their UCO in plastic canisters (*step* 1). Thereafter, staff of YLH periodically collect and transport the canisters to the processing plant, using small vans as a transport vehicle (*step* 2).
- At the plant, the UCO is mechanically cleaned and chemically converted into fatty acid methyl ester (biodiesel) through transesterification (*step 3*). The biodiesel processor used for this treatment has been developed and built by Green Fuels Ltd. (UK). Similar biodiesel processors have already been successfully implemented in other projects in developing and emerging countries (e.g., in Chiapas, Mexico, Dubai, or Hong Kong).
- The recycling product—a standard biodiesel fuel usable in all modern types of diesel engines—is sold to partnering hotels and other interested business customers as a substitute for fossil fuel (*step 4*).
- By substituting fossil fuel, the project contributes to the reduction of greenhouse gas emissions. This emission reduction has been verified by myclimate[1]

[1]http://www.myclimate.org/.

(a Swiss foundation working in the field of climate protection) according to the Gold Standard and is traded in the form of carbon-offset certificates to KUONI Travel Holding Ltd. (*step 5*).

Besides investigating the climate change mitigating effect, the project additionally aims to protect the local environment and to reduce human health risks. In Bali, UCO is commonly reused or disposed of in an unregulated way:

- The largest share of UCO is reused for food preparation. It is commonly sold via middlemen and local markets to small food stalls and households. The reuse of UCO as foodstuff causes high health risks to consumers with those affected often coming from a lower social stratum (Saguy and Dana 2003). The practice of selling UCO as foodstuff can be categorized as very hazardous as well as possibly illegal, especially as the origin of the oil is often not disclosed to customers.
- A share of the UCO is disposed of in water bodies, which increases the level of organic pollutants. Because these organic pollutants are biodegradable, aerobic degradation processes start. This leads to deteriorated water quality, thus endangering aquatic ecosystems, and—in extreme cases—may lead to septic water bodies.
- UCO disposed of on open land may infiltrate into aquifers, jeopardizing drinking water resources of the local population.
- If UCO is disposed of in unsecured landfills, it usually decomposes through anaerobic digestion processes into methane, which is about 25 times more damaging to the climate than carbon dioxide.

YLH, the social enterprise established during the course of the project, offers jobs and training to socially marginalized people from the area, allowing them to improve their living conditions and the livelihoods of their families. Further, YLH promotes the replication of its recycling solution in other areas of Indonesia, and therefore contributes to enhancing awareness and scalingup. Any financial surplus generated by the social enterprise will be reinvested into the recycling operations, as well as awareness raising campaigns and capacity-building activities.

The development phase of the project took almost three years, during which time the social enterprise and its collection and processing system have been set up. The structures and mechanisms established are intended to remain running for at least a decade.

The project approach as well as the location is rather uncommon for CaCH. Most of the CaCH's projects are implemented in lesser developed countries and have a stronger pro-poor focus. However, CaCH has considered environmental and climate issues as part and parcel of development cooperation for many years. Being already profoundly involved in climate change adaptation and disaster risk reduction programs, CaCH is currently reinforcing its climate-related activities by fostering climate change mitigation projects. Replacing fossil fuels with renewable energies, as well as the introduction of energy-efficient technologies, is in the focus of CaCH's climate change mitigation projects in developing and emerging countries. These projects consistently have strong elements of socioeconomic development and/or alleviation of poverty.

The UCO project contributes, beyond the mentioned project goals, to also increase the knowledge and expertise of CaCH in the following fields:

- Reinforcing the integration of waste recycling and renewable energy supply in development projects.
- Gaining further expertise with social business approaches as an alternative to the more common community-based development approaches.
- Gaining experience in carbon finance as a complementary finance mechanism to common donor-funded projects.

## 22.2 Design and Methods

Before project implementation began, the feasibility of the project had been thoroughly assessed. The following points have been included from the feasibility assessment:

- Important project stakeholders from the government, civilian society, and the tourism sector in Bali have been identified and their roles defined. Cooperation agreements have been made with different local government agencies and the Swiss project partners, KUONI Travel Holding Ltd., and myclimate.
- Assessment of UCO supply. More than 300 hotels and restaurants had been visited to determine UCO sources and the amount generated. Many hotels and restaurants showed high interest to collaborate, but most of them were not willing to enter cooperation agreements in the early stage of the project. The assessment clearly indicated that the amount of UCO generated in the hotels and restaurants visited is at least two times higher than the minimum amount of 1 $m^3$ per workday, which had been anticipated as a requirement for an economically viable recycling operation (cf. Sect. 22.3).
- Interested customers for the produced biodiesel have been identified. Many of the visited hotels have shown interest in buying the produced biodiesel if a high product quality, e.g., in accordance with international biodiesel standards, can be guaranteed and that the biodiesel is not more expensive than fossil diesel fuel. However, few hotels have been willing to sign preliminary purchase agreements in the early stage of the project.
- Different international manufacturers of biodiesel plants were identified and four price quotations obtained. Except for the quotation from Greenfuels Ltd. UK, all others proved to be either too expensive or insufficient.
- Legal and administrative requirements to found and operate a social enterprise in Indonesia were assessed and the organizational structure for the social enterprise developed together with the local government.

- Financial projections to verify the economic viability of the oil recycling have been developed. It has been shown that an economically viable operation is feasible if at least 1 $m^3$ of biodiesel per day could be produced. The economic calculation did not include investment costs and costs for long-term backstopping from CaCH. Investment and backstopping are covered through international funding (70 %) and revenue from emission trading (30 %).
- Project tasks had been further specified and a detailed implementation schedule developed. Initially, it was planned to finalize implementation within 16 months. However, implementation was delayed by seven months due to late import of the biodiesel processor and an extended training phase for YLH.
- A detailed budget had been developed and the project funding verified. The total implementation costs including a 10-year backstopping mandate (term of the emission trading agreement) amount to US$1.2 million. Costs are covered through funding from REPIC (an interdepartmental platform from the Swiss Government, which promotes renewable energies and energy efficiency in developing and emerging countries), the Canton of Geneva and DEG (German Investment and Development Corporation), CaCH funds, and income from the sale of emission certificates.

After one year, the project feasibility had been verified and the implementation phase started. The implementation included the following main steps:

- **Setup of the UCO processing and collection infrastructure**. A long-term lease agreement for an operations building located in a trade area in Denpasar (Capital of Bali) had been signed and the building structures adapted and renovated. The biodiesel processor had been assembled by Greenfuels Ltd. in the United Kingdom and exported to Indonesia. Bulk storage tanks, the piping system, and accessories, as well as four vans for UCO collection, had been purchased in Indonesia.
- **Foundation of the social enterprise, YLH**. The social enterprise YLH has been founded according to Indonesian law as a non-profit organization that owns and operates a private limited company. A company owned by a non-profit organization is allowed to conduct business but revenues cannot be distributed as profit. Revenues instead will be reinvested into the recycling operations and other non-profit activities. Six board members volunteered to steer and monitor YLH. The board members are not involved in daily operations. On the board, individuals with different professional backgrounds are represented. Through the board members, YLH is well connected to different stakeholder groups such as the government, civil society groups, and the tourism industry. At the operational level, eight full-time positions are required to operate and maintain collection and production. Different socially disadvantaged individuals are employed at YLH. To assure sound management, executive and key positions at operational level were filled with professionals, independent of their social situation.
- **Raising awareness of the UCO problem with hotels and restaurants, the government and the public**. One of the focus activities had been to convince as

many hotels and restaurants as possible to participate in the project. To this end, approximately 400 hotels and restaurants had been visited. Furthermore, joint workshops with the government and meetings with hotel and restaurant associations had been organized. Newspapers and a local television station also participated in some of the meetings and published reports and articles on the project. The promotion campaign not only enhanced public interest for the project but also garnered higher interest from hotels and restaurants in participating. Up until December 2013, the collection rates were continuously increasing, but are still considered too low for a cost-effective recycling (Sect. 22.3).

- **Sales agreements for the produced biodiesel had only been established after production started**. The sales price is about US$1 per liter, approximately equivalent to the price of non-subsidized fossil diesel fuel. The produced biodiesel was initially used for emergency generators and hot water boilers at hotels and construction equipment of a building contractor.
- **Certification of emission trading according to the Gold Standard had been prepared by myclimate**.
- **Agreement for long-term backstopping for YLH**. For various reasons, CaCH had agreed to continue the support of YLH in the long-term: (1) The business and management capacity of YLH is still limited and requires support; (2) Regular monitoring for verification of the carbon offsetting volume is required; and (3) Keeping track of the development of the UCO recycling also provides an important learning experience for CaCH.

## 22.3 Results

The project's success hinged to a large extent on the economies of scale and, therefore, on the amount of UCO available for recycling. The more the biodiesel is produced, the better the cost/income ratio. At project start, economic projections have shown that at least 1 $m^3$ of UCO per workday (21 $m^3$/month) will have to be recycled to assure cost-covering recycling operations. These initial projections had been based on assumptions and technical references.

In December 2012, the project started gradually with UCO collection. In March 2013, a normal operation mode had been established. Figure 22.2 shows the volume of UCO collected per month between March and November 2013. The collection rate showed a very positive increasing trend, but with 13 $m^3$ in November 2013, it is still below the minimum amount expected and necessary for cost coverage. At that time, the number of hotels and restaurants regularly providing UCO to YLH amounted to 122.

An economic viability calculation had been carried out for handing over the recycling operations to YLH in July 2013. At that time, the social enterprise was still far away from having a cost-covering operation because of the low amount of

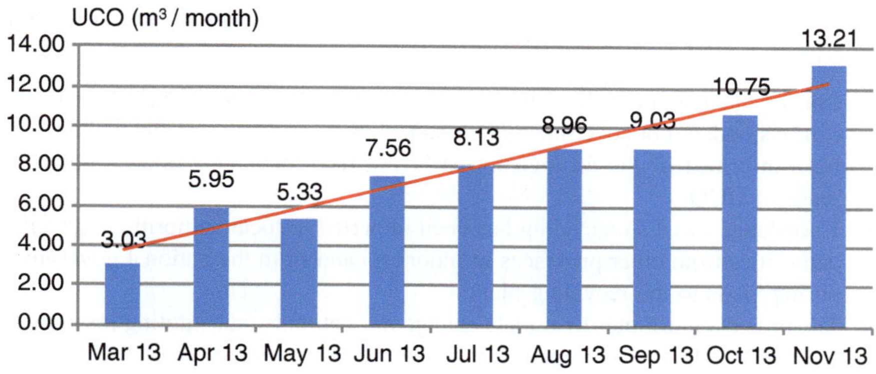


**Fig. 22.2** Monthly volume of UCO collected between March and November 2013

UCO collected. Only 8 $m^3$ had been collected in July 2013. Yet, the collection rate already showed signs of growth with the amount of UCO collected increasing from roughly 0.9–1 $m^3$ each month. Through this development, losses decreased month to month. Based on this trend, profit and losses were calculated and projected for the period from July 2013 to July 2014. Assuming the trend held, YLH will be cost-covering from July 2014 onwards (cf. Fig. 22.3). The calculation further indicated that a minimum amount of approximately 19 $m^3$ per month is required to effectively cover costs. This is roughly equal to a daily amount of 0.9 $m^3$ for five workdays per week. The calculation generally affirmed the projection (21 $m^3$/month) from the project start. For a period of 13 months, till YLH was forecasted to break even in July 2014, losses of about US$35,000 were incurred and were covered through CaCH funds.

**Project Achievements**. If the financial projections prove to be accurate, the project has the potential to become a best practice example for UCO collection in

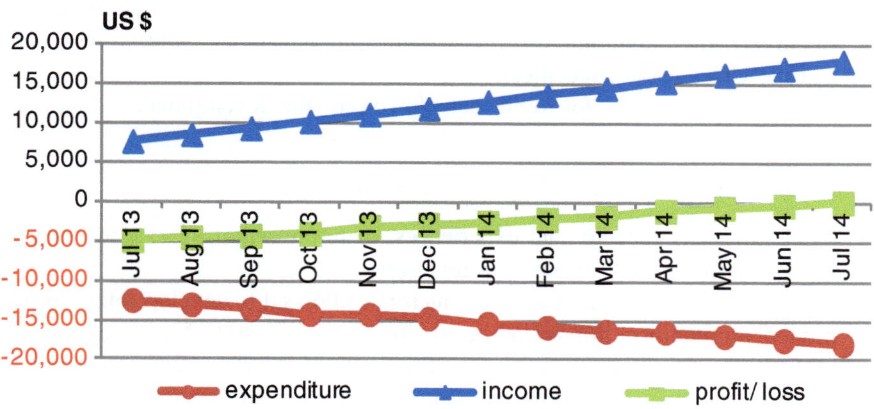


**Fig. 22.3** Profit/loss projection for biodiesel production

emerging market countries. The integrated approach provides impact at different levels:

- Local stakeholders such as hotels, restaurants, the government, tourism associations, other nongovernmental organizations (NGOs), and the public have been informed of the ecological and health risks of inappropriate disposal or reuse of UCO.
- Knowledge of UCO recycling has been imparted to local authorities, as well as authorities from other provinces in Indonesia and from the national government during visits to the recycling plant.
- The project contributes to climate change mitigation by substituting fossil diesel fuel with the biodiesel produced. It is expected that a minimum of 1000 tones of $CO_2$ emissions will be avoided each year.
- Training and income opportunities for socially disadvantaged people have been created.
- For CaCH, this pilot project provides new knowledge and experience with regard to subjects such as renewable energies, carbon finance mechanisms, closing material flows (circular economy), and social business approaches.

**Project Obstacles**. Besides the promising results so far, there are several factors which impair project effectiveness. Some of the obstacles are caused by project-external factors that normally require great effort and resources to be changed. The following obstacles are related to external factors:

- Environmental awareness and social responsibility in the tourism sector are increasing but still low. Many stakeholders use this issue for public relations measures but in most cases, understanding and commitment are still limited.
- No legal framework exists that facilitates an environmentally sound and healthy recycling of UCO.
- Only limited interest, capacity and thus support from government stakeholders.
- No legal framework to promote social business activities and the cooperation between municipalities and social entrepreneurs. In actuality, this could be one option to improve public services and also the capacity of local authorities.
- Relatively low public interest in environmental and social subjects, in general, and the UCO subject specifically.
- Government subsidies for fossil fuels do confine the development of renewable energy solutions.

Other obstacles are related to project-internal factors:

- YLH still lacks capacity and experiences, therefore requiring external support, especially in the field of business management.
- The organizational form of a social enterprise allows it to develop solutions with limited profitability, not yet attractive for real businesses. Moreover, the low profitability also limits the capacity of the social business. Average salaries for management positions are not attractive to experienced managers. The experience provided by the board members can compensate for this weakness in part,

but since board members are working on a voluntary basis, their involvement is limited.

- Only finite financial resources for long-term backstopping from CaCH are available. It has been highly compelling for donors to support the project implementation. Yet, it is still difficult to find funding for continued support and capacity building of YLH.
- CaCH as a NGO has limited influence on government policies and their legal framework. If it would be possible to mobilize more support from government agencies, it would likely also be easier to convince hotels and restaurants to participate on in the UCO recycling.

## 22.4 Conclusions

Authorities (public economy) are usually responsible for public service provision (Robinson and White 1997). Assuming ideal conditions exist, state solutions are the first-best solution to solve social problems, since the state does not possess any self-interest. Developing and emerging countries often show large deficits in their institutional framework, thus no ideal conditions are prevalent. Options to circumvent these shortcomings can be provided by non-profit organizations (social economy), companies (private profit-oriented economy), or social enterprises. The latter can be perceived as a hybrid of the two former. Therefore, non-profit organizations and for-profit companies may be considered functional equivalents to the state in solving social problems. Collaborative arrangements between the different forms have further potential to create synergies (Helling et al. 2005). The only remaining question is under what conditions which approach proves to be the most practical instrument (Beckmann 2011).

In the case of UCO recycling in Bali, the following can be concluded: Based on weak state institutions and a lack of legislative framework, no state-based solutions to the UCO problem have been developed thus far. Further, high economic risks and the required start-up capital have prevented the development of formal private sector solutions for UCO recycling. As a result, the social enterprise solution constitutes a second-best solution. The social enterprise can be perceived as a kind of transformation agent to overcome the existing problem. This involves bridging the institutional problems of the state system until a first-best solution is feasible. With this in mind, public service provision by social enterprises can be understood as a temporary solution until the state is able to meet its obligations (Beckmann 2011).

Looking at the project in the context of a second-best solution, the following lessons have been learned: Local authorities should have been more strongly involved and obligated to provide long-term and active support to the social enterprise. A larger role by authorities would have increased the pressure on hotels and restaurants to participate in the UCO recycling. Further, the social enterprise also requires that after the implementation phase, a more systematic form of support from external groups is available to increase awareness of the UCO problem and to improve the knowledge and capacity of important stakeholders, such as hotels and restaurants, tourism associations, different government agencies, and the public. Additionally, the possibility of financial support should also be ensured beyond the implementation phase, until such time that the UCO recycling has become established in an economically sustainable way. According to Helling et al. (2005), the mentioned points can be summarized as follows:

- Improve the enabling environment through support from the government and civil society.
- Enhance local capacities through social capital development, improvement of knowledge, and skills.
- Ensure support through finance means, technical assistance, and information.

During project implementation in Bali, these points have been considered. However, they were not a main focus. Hence, in the case of replication of the project, a higher prioritization of these points should be considered.

## References

Beckmann, M. (2011). Social entrepreneurship. Neues Phänomen, altes Paradigma oder Vorbote eines Kapitalismus 2.0. In S. Empter & H. Hackenberg (Ed.), *Social entrepreneurship—Social business: Für die Gesellschaft unternehmen*. Wiesbaden: Verlag (pp. 67–85).

Helling, L., Serrano, R., & Warren, D. (2005). Linking community empowerment, decentralized governance, and public service provision through a local development framework. *Social Protection Discussion Paper No. 535*. Washington, DC: World Bank.

Robinson, M., & White, G. (1997). The role of civic organizations in the provision of social services. Towards synergy. *Research for Action, 37*. Helsinki: UNU World Institute for Development Economics Research.

Saguy, I. S., & Dana, D. (2003). Integrated approach to deep fat frying: Engineering, nutrition, health and consumer aspects. *Journal of Food Engineering, 56*(2–3), 143–152.

MIX
Papier aus verantwortungsvollen Quellen
Paper from responsible sources
FSC® C105338

If you have any concerns about our products,
you can contact us on
**ProductSafety@springernature.com**

In case Publisher is established outside the EU,
the EU authorized representative is:
**Springer Nature Customer Service Center GmbH**
**Europaplatz 3, 69115 Heidelberg, Germany**

Printed by Libri Plureos GmbH
in Hamburg, Germany